ROLLS-ROYCE HERITAGE TRUST

FLOW MATCHING OF THE STAGES OF AXIAL COMPRESSORS

Geoffrey Wilde OBE

TECHNICAL SERIES No 4

Published in 1999 by the
Rolls-Royce Heritage Trust
P O Box 31 Derby England DE24 8BJ

© 1998 Geoffrey Wilde OBE

This book, or any parts thereof, must not be reproduced in any form without the written permission of the publishers.

ISBN: 1 872922 14 7

Previous volumes published in this Technical Series, and the Rolls-Royce Heritage Trust Historical Series, are listed at the rear.

Cover Picture: Cross-section of the variable stator compressor designed for the Rolls-Royce Avon RA3 engine (BT Sch 4011). This option was not used.

Books are available from:
Rolls-Royce Heritage Trust, Rolls-Royce plc, Moor Lane, PO Box 31, Derby DE24 8BJ

Origination and Reproduction by Neartone Ltd, Arnold, Nottingham
Printed by Premier Print, Glaisdale Parkway, Bilborough, Nottingham

CONTENTS

Page

Foreword		5
Chapter One	Introduction and objective	7
Chapter Two	Axial compressor performance with variation of rotational speed	11
Chapter Three	Axial compressor stage flow mismatch at speeds below design speed	19
Chapter Four	Performance of compressor aerofoils	27
Chapter Five	The design problem	31
Chapter Six	The pressure ratio - flow characteristics of axial compressors	37
Chapter Seven	Examples of axial compressors in practice	53
Chapter Eight	Summary	74

Fig 1 Rolls-Royce Avon RA24 (200 Series) axial turbojet

FOREWORD

The flow matching of the stages in a high pressure ratio axial compressor operating in a gas turbine engine is a complex and difficult task. This was demonstrated more than fifty years ago when the Company designed its first axial turbojet engine, the Rolls-Royce AJ65. Although over the years, giant strides have been made at the engine design point with improved design tools and methods (such as the powerful modern computers and computational fluid dynamics {CFD}), away from the design point, flow matching is still difficult and largely empirical. Forty years after the AJ65, in the 1980s when the Company was involved in the design of a new core compressor for the V2500 turbofan, one of the major development problems was in matching the compressor stages to achieve the required performance over the entire engine operating envelope. The problem was solved by the fundamental understanding and empirical rules established over the years from many hours of rig and engine testing.

This paper attempts to set out the fundamentals and principles of this understanding. It was written by Geoffrey Wilde, after he had retired from the Company, as a design exercise for his students at the Universities of Liverpool and Loughborough where he was a visiting Industrial Professor. I do not believe there is anybody better qualified or who has done more to further our understanding than Geoffrey Wilde. He joined the Company in September 1938 and spent his first year working for Alf Arnold on the Experimental Test Beds gaining an understanding of piston engine performance. He then went to work for Stanley Hooker to help in the development of engine superchargers, taking over the Supercharger Performance Section in 1943 when Hooker went to Barnoldswick. His piston engine work on the Merlin and Griffon led to his design of the compressors for the Dart turboprop engine and the formation of the Compressor Aero Design Office. By this time, the Company's first axial turbojet, AJ65, was in serious trouble with compressor surge. It was transferred from Barnoldswick to Derby. Geoffrey initiated the principle of variable angle stator blading linked to progressively variable interstage bleed to postpone the onset of blading stall. This was successful in enabling the front stages of the compressor to run stably, even when operating a long way from the compressor design point. This was the first time that this method had been applied to enable high pressure ratio axial compressors to operate reliably and safely at off-design conditions. The resulting engine became the well known Avon axial jet which was used extensively in most of the Royal Air Force's front line fighters and bombers of the 1950s and 60s. He went on to design the compressors for the Conway engine, the next stage in the development of axial compressors, before he left the Compressor Office to lead the Advanced Projects Office in 1956. He remained closely associated with all the Company's work on its advanced projects until he retired in 1979. He will perhaps be best remembered for his work in initiating the hollow wide chord fan blade and the three-shaft engine. These concepts have enabled the Trent engine to take a dominant share of today's market. Geoffrey Wilde's concepts are thus of growing importance.

Dave Piggott
March 1999

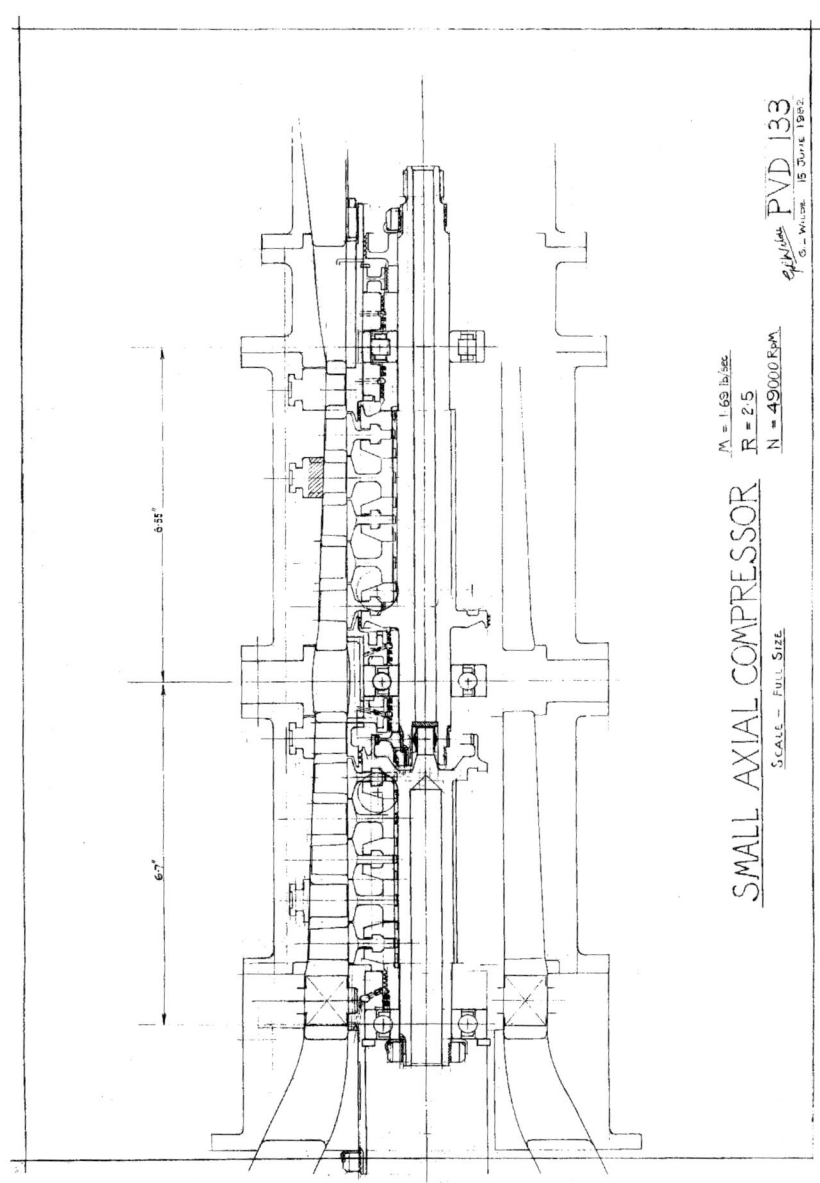

Fig 2 Cross-section of a small axial compressor used as a supercharger for a racing car engine. Note the blade chords are relatively large for performance and mechanical reasons.

FLOW MATCHING OF THE STAGES OF AXIAL COMPRESSORS

CHAPTER ONE

Introduction and objective

The gas turbine engine, and particularly the aero gas turbine, has to function smoothly and safely over a wide range of engine rotational speeds and a wide range of aircraft flight speeds and conditions.

It is a relatively straightforward task to design an axial compressor to operate at a given condition. This is usually referred to as the aerodynamic design point. A typical condition could be the engine rotational speed required for the engine to give the power necessary to climb the aircraft from the runway to the cruising altitude. But equally important is the capability of the engine and its compressor to start, accelerate quickly to maximum power, and to accept passively being as quickly shut down, all with smooth functioning of the components without causing excessively high stresses and temperatures within the engine.

Axial compressors are well known for their tendency to generate pressure oscillations known as 'surge' when the air flow rate passing through the compressor falls below a critical value referred to as the surge point. This surge point is related to the operating conditions of the individual stages of the compressor, and particularly to the angle at which the air approaches the aerofoils. Above a certain angle the aerofoils stall in a similar manner to the wings of an aircraft and it is the responsibility of the compressor designer to ensure that the front stages of the compressor in particular do not stall in any important operating condition of the engine.

The Rolls-Royce Avon 200 Series engine *(Fig 1)* embodying an axial compressor delivered a thrust of 10000 lbf (44.48 kN) when run at full speed at the start of takeoff on the runway. It was used extensively in the Royal Air Force's front line fighter and bomber aircraft in the 1950s and early 1960s. The compressor had 15 stages, generated a pressure ratio of 10:1, and delivered 150 lb/sec of air (68 kg/sec). The inlet diameter of the first stage rotor was 32 inches (0.81 metres). Design features were incorporated to guard against the danger of running into compressor surge.

In contrast, the very small axial compressor *(Fig 2)* was designed as part of a compound supercharging system for a high performance piston engine for a racing car. It had 6 stages, generated a pressure ratio of 3.5:1 and delivered 1.75 lb/sec of air (0.79 kg/sec). The inlet diameter of the first stage was 4.7 inches (0.12 metres). At this much lower pressure ratio there was not any need to incorporate special design features to guard against compressor surge because this was unlikely to occur, the flow matching of the stages at lower speeds being such that the first and second stages would not be forced to running in a stalled state.

Another example of a small simple axial turbojet engine is the Armstrong Siddeley Viper *(Fig 3)* which was used in trainer aircraft such as the Jet Provost in the RAF in 1960s.

The Viper gave a thrust of 2500 lbf (11.12 kN) at sea level static conditions. The compressor had 7 stages, generated a pressure ratio of 4:1 and delivered 44 lb/sec of air (20 kg/sec). The inlet diameter of the first stage rotor was 18.5 inches (0.47 metres). The rotor assembly *(Fig 4)* and the two half casing of stator blading *(Fig 5)* are typical of this class of small engine. Again, at this modest pressure ratio, no design provision was necessary to guard against compressor stall. The Viper could be called the axial compressor equivalent of Whittle's jet engine which had a design pressure ratio within the range 4 to 4.5.

The object of this paper is to introduce the engineering student to the problems of the flow matching of the stages in an axial compressor to ensure smooth and stable operation under all operating conditions.

These problems in reality are technically very complex. They have been the subject of much theory and experiment over some fifty years since the first gas turbine engines with axial compressors were built and installed in aircraft.

To introduce these problems, it is essential to make drastic simplifications of the technical factors involved yet retain the fundamental performance characteristics that can cause unstable running of engines.

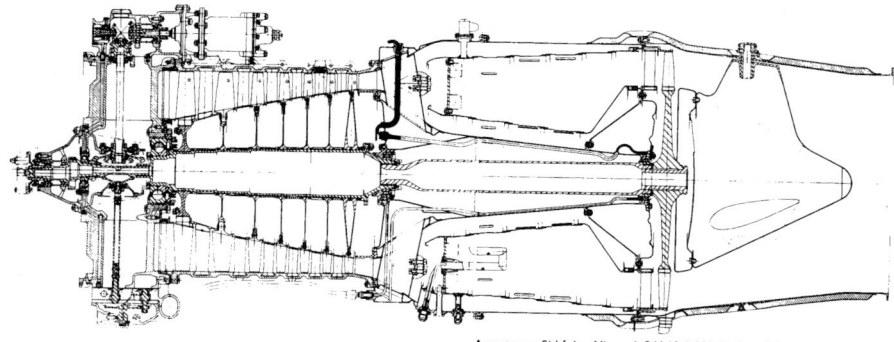

Fig 3 Cross-section of Armstrong Siddeley Viper axial compressor

This can be done and it is thought that the methods proposed here will give the student practical insight into this important subject while at the same time affording him the opportunity of suggesting design solutions.

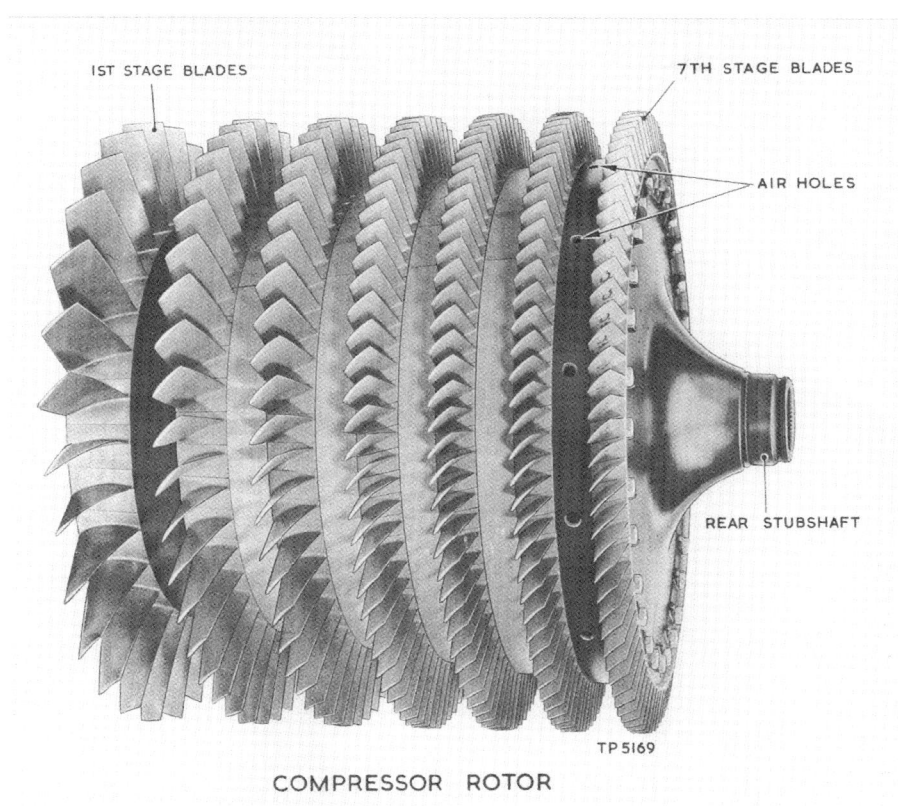

Fig 4 The rotor assembly of Armstrong Siddley Viper turbojet

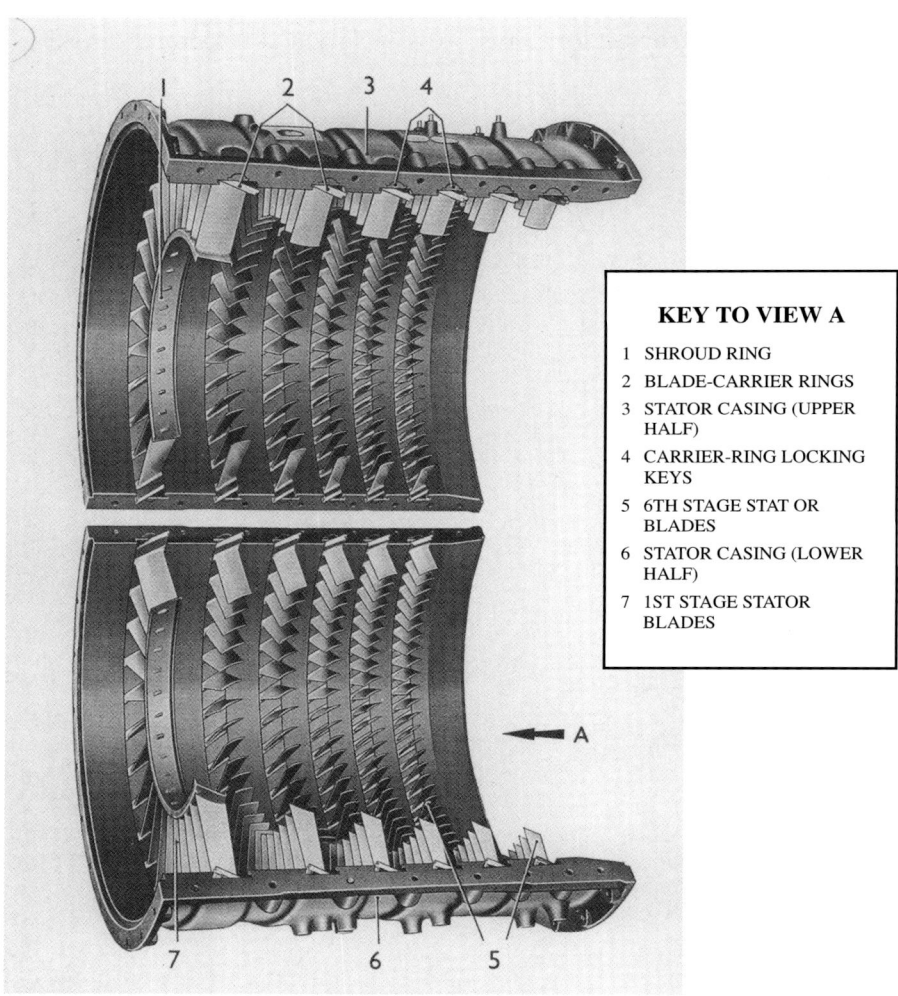

Fig 5 The stator assembly of Armstrong Siddeley Viper turbojet

CHAPTER TWO

Axial compressor performance with variation of rotational speed

An axial compressor consists of a series of rotating discs to which are attached a large number of curved plates usually referred to as cambered aerofoils *(Fig 6 upper)*. They are set in such a way that at the aerodynamic design point of the compressor the air approaches them at substantially the same angle as that of the forward edge of the aerofoils and thus with the minimum of directional shock to the flow.

As the air passes through the cambered aerofoils *(Fig 6 lower)* it is given an increase in tangential velocity (Vw_3–Vw_4) which increases the total energy of the air. This produces an increase in static pressure in the air stream and also an increase in velocity and kinetic energy in the air stream (from V_4 to V_3). Behind each rotating disc of cambered aerofoils there is a stationary disc of cambered aerofoils, the object of which is to receive the higher velocity air stream from the rotating aerofoils without directional shock and to reduce the velocity of this stream (V_3 to V_4) further raising the static pressure in the air stream.

This process is repeated through each pair of rotating and static rows of aerofoils. Each pair is called an axial compressor stage and there may be as many as 12 stages in series to produce a delivery pressure of 6 times the compressor inlet pressure when pumping air.

Fig 7 is a cross-section of the Rolls-Royce Spey turbofan engine. It has two compressors in series running at different rotational speeds and each driven by two turbine stages. The first compressor (low pressure – LP) produces a pressure ratio of 3:1 and the second compressor (high pressure – HP) produces 6:1 giving 18:1 overall.

At entry to each compressor stage the product of the flow annulus area (A), air density (ρ), and axial velocity (Va) (sometimes called the 'flux' or 'through flow') is constant and equal to the mass flow rate (M) of the air passing through the compressor. As the pressure and density of the air increases through the stages, the flow annulus area must progressively decrease through the stages and the radial length of the curved plates or cambered aerofoils diminishes correspondingly as shown in *Fig 7*. In the case of the HP compressor in *Fig 7* the axial velocity through the stages at the chosen design condition is constant. This together with the chosen constant mean radius of the truncated annular flow passage giving the same mean blade speed for all the stages is a convenient simplification of aerodynamic design which is very suitable for this paper.

We will now consider five compressors working on air with overall

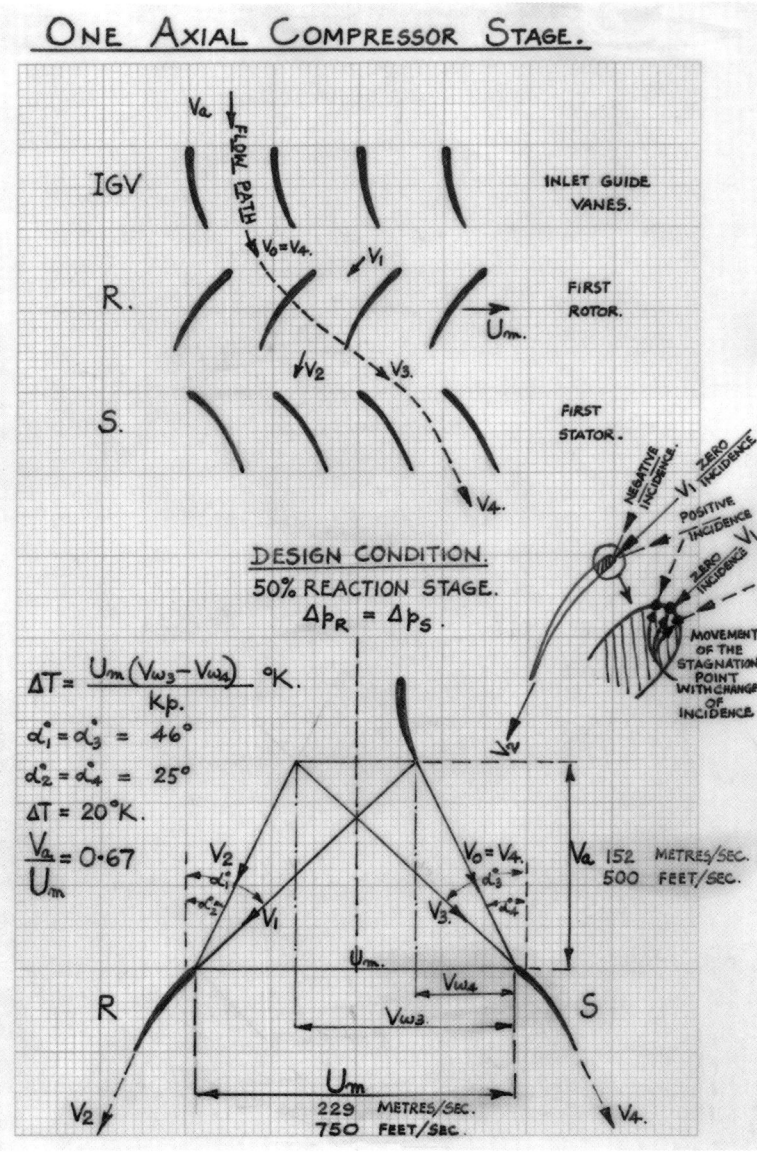

Fig 6 Velocity triangles of one axial compressor stage

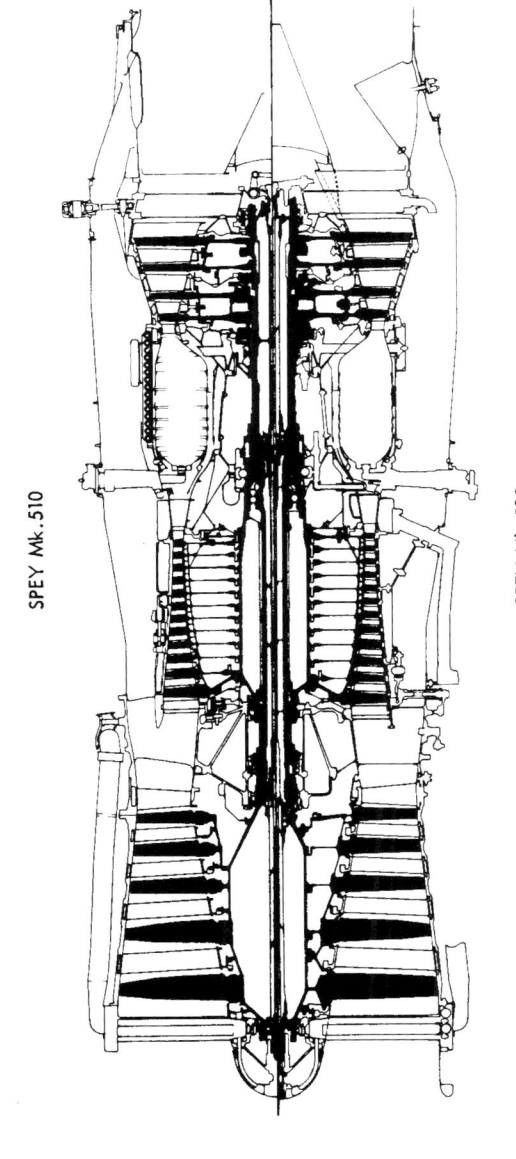

Fig 7 Cross-section of Rolls-Royce Spey turbofan showing a comparison of the civil Mark 510 and the military Mark 202 engines

pressure ratios (P_2/P_1) of 3:1; 6:1; 10:1; 15:1; and 20:1 respectively, all having the same through flow axial velocity (Va) at the design condition and the same flow annulus mean radius (Rm). Hence the mean tangential speed (Um) of all the cambered aerofoils is the same for all stages.

The following simplifying assumptions are made:

(a) Axial velocity (Va) through all stages at the design condition is constant. Typical value of Va = 500 ft/sec (152.4 metres/sec).

(b) Mean tangential speed (Um) of all cambered aerofoils is constant at the flow annulus mean radius (Rm). Typical value of Um = 750 ft/sec (228.6 metres/sec).

(c) For air $\gamma = 1.4$ constant

$$\therefore \frac{(\gamma - 1)}{\gamma} = 0.2857$$

(d) Adiabatic efficiency of each stage $\eta_{ad} = 85.72\%$ treated the same as polytropic efficiency η_p and assumed constant at all conditions – a considerable approximation only acceptable for the purposes of this paper.

(e) Inlet temperature to compressor $T_1 = 288°K$.

(f) Compressor outlet/inlet overall temperature ratio T_2/T_1.

$$\frac{(T_2)}{(T_1)} = \frac{(P_2)}{(P_1)}^{(1/\eta_p)(\gamma-1)/\gamma} = \frac{(P_2)}{(P_1)}^{1/3}$$

(g) Compressor outlet/inlet air density ratio (ρ_2/ρ_1)

$$\frac{\rho_2}{\rho_1} = \frac{P_2}{P_1} \cdot \frac{T_1}{T_2} = \frac{(P_2)}{(P_1)}^{2/3}$$

(h) Air mass flow rate M

at inlet	$M = A_1 Va_1 \rho_1$
at outlet	$M = A_2 Va_2 \rho_2$
at design condition	$Va_1 = Va_2 = Va_m$

$$\therefore \quad \frac{A_1}{A_2} = \frac{\rho_2}{\rho_1} = \frac{(P_2)^{2/3}}{(P_1)}$$

(i) Compressor overall temperature rise $(T_2-T_1) \propto Um^2$.

This is valid for the middle stages of the compressor where Va/Um is substantially constant as Um varies (velocity triangles scale), but is in error for the low pressure stage at the front of the compressor where Va_1/Um decreases as compressor speed is reduced, and also in error for the high pressure stage at the rear of the compressor where Va_2/Um increases as compressor speed is reduced.

However the combination of these opposing effects at the front and rear of the compressor tend in practice roughly to cancel out and the assumption of

$$(T_2 - T_1) \propto Um^2$$

is acceptable as an approximation for the purposes of this design exercise.

The design condition chosen for an axial compressor is commonly at about 90% of the maximum allowable aerodynamic speed. This maximum speed is usually the speed above which the efficiency of the compressor falls rapidly due to the air impinging onto the blading at a velocity approaching 80 to 85% of the velocity of sound in the air. The ratio of the impinging velocity (V_1) to the velocity of sound (Vs) is called the Mach Number (No) which is an important physical factor in the design of turbomachinery.

The following table *(Fig 8)* gives the variations in temperature rise through the compressor, temperature ratio, density ratio, and pressure ratio for the following fractions of the design speed 25%, 50%, 75%, 100% and 110%. The density ratio is plotted against the fraction of design speed in *Fig 9*.

The compressor truncated annular flow passages for the five compressors of increasing pressure ratio are also illustrated in *Fig 9*.

As already noted the compressor flow annulus dimensions were determined in each case by prescribing Va/Um constant through all the compressor stages at the chosen speed only (i.e. at $N_D = 100\%$ in *Fig 8*).

At speeds away from design the ratio of axial velocity Va to mean blade speed Um remains substantially constant in the middle of the compressor, but departs markedly from the design value in the front and rear stages of the compressor. This is considered in the next chapter. In order to limit the computational task the operating conditions of the first and last stages only of the compressors are considered in this paper.

COMPRESSOR OVERALL DENSITY RATIO VARIATION WITH SPEED (RPM)

DESIGN PRESSURE RATIO $Rc = P_2/P_1 = 3.00$					DESIGN PRESSURE RATIO $P_2/P_1 = 6.00$					DESIGN PRESSURE RATIO $P_2/P_1 = 10.00$					DESIGN PRESSURE RATIO $P_2/P_1 = 15.00$					DESIGN PRESSURE RATIO $P_2/P_1 = 20.00$				
$\%N_D$	ΔT °K	$\frac{T_2}{T_1}$	$\frac{P_2}{P_1}$	$\frac{\rho_2}{\rho_1}$	$\%N_D$	ΔT °K	$\frac{T_2}{T_1}$	$\frac{P_2}{P_1}$	$\frac{\rho_2}{\rho_1}$	$\%N_D$	ΔT °K	$\frac{T_2}{T_1}$	$\frac{P_2}{P_1}$	$\frac{\rho_2}{\rho_1}$	$\%N_D$	ΔT °K	$\frac{T_2}{T_1}$	$\frac{P_2}{P_1}$	$\frac{\rho_2}{\rho_1}$	$\%N_D$	ΔT °K	$\frac{T_2}{T_1}$	$\frac{P_2}{P_1}$	$\frac{\rho_2}{\rho_1}$
110%	154	1.54	3.62	2.36	110%	285	1.99	7.86	3.95	110%	402	2.40	13.76	5.74	110%	511	2.77	21.30	7.69	110%	597	3.07	29.10	9.45
100%	127.4	1.44	3.00	2.08	100%	235	1.82	6.0	3.30	100%	332	2.15	10.00	4.64	100%	422	2.47	15.00	6.08	100%	494	2.71	20.00	7.36
75%	7.16	1.25	1.95	1.56	75%	132	1.46	3.10	2.13	75%	187	1.65	4.49	2.72	75%	238	1.82	6.07	3.33	75%	278	1.96	7.58	3.86
50%	31.8	1.11	1.37	1.23	50%	59	1.20	1.75	1.45	50%	83	1.29	2.14	1.66	50%	106	1.37	2.55	1.87	50%	123	1.43	2.91	2.04
25%	7.96	1.03	1.08	1.06	25%	1.47	1.05	1.16	1.10	25%	21	1.07	1.23	1.15	25%	26.4	1.09	1.30	1.19	25%	30.9	1.11	1.36	1.23

N_D = Design speed T_1 = Inlet temperature = 288°K T_2 = Outlet temperature °K

ΔT = Overall temperature rise °K = $(T_2 - T_1)$ ρ_2/ρ_1 = Outlet to inlet air density ratio

Fig 8 Table of compressor overall density ratio variation with speed (rpm)

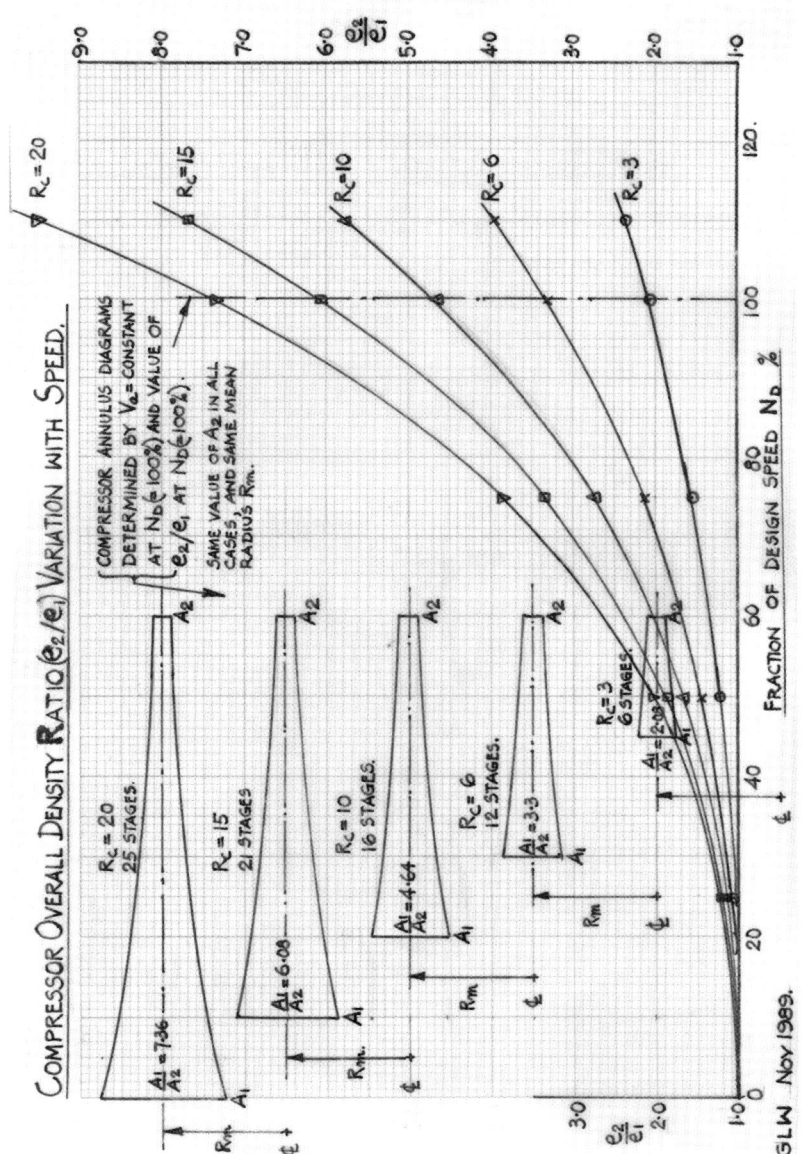

Fig 9 Compressor overall density ratio (ρ_2/ρ_1) variation with speed

FIRST AND LAST STAGE AXIAL VELOCITIES WITH VARIATION OF COMPRESSOR SPEED

DESIGN PRESSURE RATIO $Rc = P_2/P_1 = 3.00$ $A_1/A_2 = 2.08$				
$\%N_D$	$\frac{\rho_2}{\rho_1}$	$\frac{Va_1}{Va_2}$	$\frac{Va_1}{Um}$	$\frac{Va_2}{Um}$
110%	2.36	1.13	0.71	0.63
100%	2.08	1.00	0.67	0.67
75%	1.56	0.75	0.58	0.77
50%	1.23	0.59	0.51	0.87
25%	1.06	0.51	0.48	0.94

DESIGN PRESSURE RATIO $P_2/P_1 = 6.00$ $A_1/A_2 = 3.30$				
$\%N_D$	$\frac{\rho_2}{\rho_1}$	$\frac{Va_1}{Va_2}$	$\frac{Va_1}{Um}$	$\frac{Va_2}{Um}$
110%	3.95	1.20	0.73	0.61
100%	3.30	1.00	0.67	0.67
75%	2.13	0.65	0.54	0.83
50%	1.45	0.44	0.44	1.01
25%	1.10	0.33	0.38	1.17

DESIGN PRESSURE RATIO $P_2/P_1 = 10.00$ $A_1/A_2 = 4.64$				
$\%N_D$	$\frac{\rho_2}{\rho_1}$	$\frac{Va_1}{Va_2}$	$\frac{Va_1}{Um}$	$\frac{Va_2}{Um}$
110%	5.74	1.24	0.75	0.60
100%	4.64	1.00	0.67	0.67
75%	2.72	0.59	0.51	0.87
50%	1.66	0.36	0.40	1.12
25%	1.15	0.25	0.34	1.34

DESIGN PRESSURE RATIO $P_2/P_1 = 15.00$ $A_1/A_2 = 6.08$				
$\%N_D$	$\frac{\rho_2}{\rho_1}$	$\frac{Va_1}{Va_2}$	$\frac{Va_1}{Um}$	$\frac{Va_2}{Um}$
110%	7.69	1.26	0.75	0.59
100%	6.08	1.00	0.67	0.67
75%	3.33	0.55	0.50	0.91
50%	1.87	0.31	0.37	1.21
25%	1.19	0.20	0.30	1.15

DESIGN PRESSURE RATIO $P_2/P_1 = 20.00$ $A_1/A_2 = 7.36$				
$\%N_D$	$\frac{\rho_2}{\rho_1}$	$\frac{Va_1}{Va_2}$	$\frac{Va_1}{Um}$	$\frac{Va_2}{Um}$
110%	9.45	1.28	0.76	0.59
100%	7.36	1.00	0.67	0.67
75%	3.86	0.52	0.48	0.93
50%	2.04	0.28	0.35	1.27
25%	1.23	0.17	0.27	1.64

A_1 = Inlet annulus area to first stage

A_2 = Outlet annulus area from last stage

Va_1 = Axial velocity into first stage

Va_2 = Axial velocity from last stage

Fig 10 Table of first and last stage axial velocities with variation of compressor speed

CHAPTER THREE

Axial compressor stage flow mismatch at speeds below design speed

The mass flow rate (M) continuity equation (cf Chapter 2 (h)) still applies at all speeds below design speed, but as the annulus dimensions have been fixed for the design case, A_2/A_1 is constant.

$$M = A_1 \cdot Va_1 \cdot \rho_1 = A_2 \cdot Va_2 \cdot \rho_2 \text{ lb/sec or kg/sec}$$

$$\therefore \quad \frac{Va_1}{Va_2} = \frac{A_2}{A_1} \cdot \frac{\rho_2}{\rho_1}$$

In order to calculate the compressor inlet (Va_1) and outlet (Va_2) velocities at speeds below design it is convenient to relate these to the stages in the middle of the compressor which operate in practice not far way from a constant value of Va/Um (cf Chapter 2 (i)).

This is done here by writing $\sqrt{Va_1 \cdot Va_2} = Va_m$ as the geometric mean of compressor inlet and outlet velocities, which is an approximation.

Therefore:

$$\frac{\sqrt{Va_1 \cdot Va_2}}{Um} = \frac{500}{750} = 0.67 \text{ design value}$$

(cf Chapter 2 (a) & (b))

from which :

$$\frac{Va_1}{Um} \cdot \sqrt{\frac{Va_2}{Va_1}} = 0.67$$

and

$$\frac{Va_2}{Um} \cdot \sqrt{\frac{Va_1}{Va_2}} = 0.67$$

The table *(Fig 10)*, using the data in *Fig 8*, gives the values of Va_1/Um and Va_2/Um from 1.1 N_D down to 0.25 N_D.

The values of Va_1/Um and Va_2/Um are plotted against % N_D in *Fig 11*.

At 50% N_D the first stage is forced to operate at $Va_1/Um = 0.37$ for the 15:1 pressure ratio compressor compared with $Va_1/Um = 0.67$ at the design speed 100% N_D. On the other hand the last stage is forced to operate at $Va_2/Um = 1.21$ compared with $Va_2/Um = 0.67$ at 100% N_D.

Fig 11 shows how increasing the design pressure ratio of the compressor from $Rc = 3$ to $Rc = 20$ forces the first and last stages of the compressor to operate progressively further away from the design condition at the lower speeds. Eventually this drives the first stage into stall in which condition the flow is unstable and may reverse, and the last stage into negative incidence stall in which condition there is a further reduction in pressure rise across the stage due to the increase in total pressure loss caused by flow separation and turbulence. In the extreme case the last stages of the compressor may be forced to 'turbine', that is, to deliver power to the rotor due to flow expansion across the stage instead of flow compression with consequent further reduction in delivery pressure. In the case considered here the last stage is, in fact, 'turbining'.

Fig 11 also includes velocity triangles for the 15:1 pressure ratio compressor at (A); (B); (C) and (D); as follows:

(A) The design condition (cf also *Fig 6*)

(B) Half speed (50% N_D) appropriate to the middle stages of the compressor since Va_m/Um is substantially constant.

(C) The first stage at 50% N_D at which $Va_1/Um = 0.37$ (cf *Fig 10*).

(D) The last stage at 50% N_D at which $Va_2/Um = 1.21$ (cf *Fig 10*).

To simplify the analysis the air outlet angles $\alpha°_2$ and $\alpha°_4$ from the blading are assumed constant as Va/Um varies. This would not be the case in practice as $\alpha°_2$ and $\alpha°_4$ would increase a little as Va_1/Um decreases.

Velocity triangle (C) shows that at $Va_1/Um = 0.37$ the increase in tangential velocity given to the air (Vw_3-Vw_4) has increased considerably compared with that at $Va_1/Um = 0.67$, and that there is a large increase in the air inlet angle $\alpha°_1$ to 66° (cf *Fig 6*) causing the aerofoil to try to operate at a high positive incidence angle of 20°. At this incidence the aerofoil will be stalled.

Similarly velocity triangle (D) shows that at $Va_2/Um = 1.21$ the reduction in tangential velocity given to the air (Vw_3-Vw_4) has reduced considerably and has actually become negative. So this stage is 'turbining'. There is a large decrease in the air inlet angle of $\alpha°_1$ to 19.8° causing the aerofoil to operate at a high negative incidence angle of -26° which would be certain to

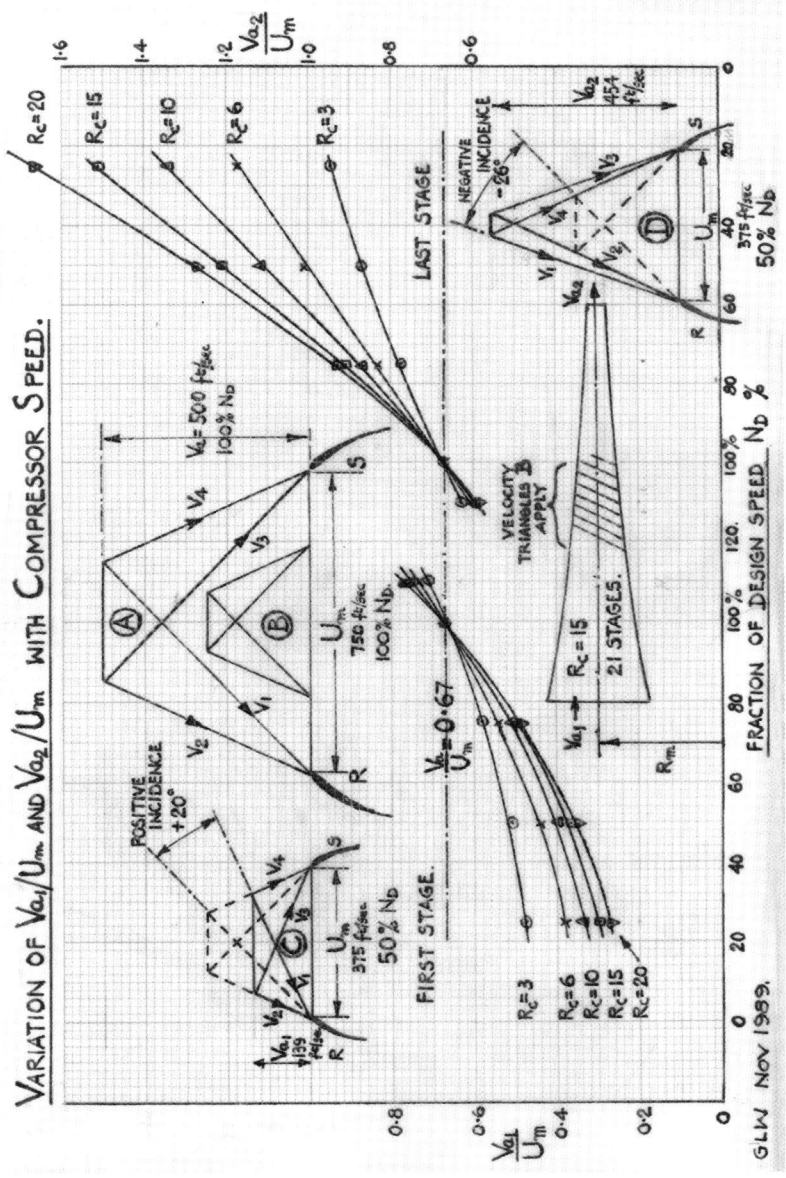

Fig 11 Variation of V_{a_1}/U_m and V_{a_2}/U_m with compressor speed

drive the stage into negative incidence stall and possibly into the choking condition. In this state the flow approaches the velocity of sound within the aerofoil passages, a dangerous condition, which can cause the aerofoils to vibrate in bending with high amplitude and stress likely to lead to mechanical failure.

Nevertheless it is the stalling of the stages at the front of the compressor that is the most damaging effect to both the performance of the engine and the mechanical reliability of the compressor blading. This is because when the longer aerofoils stall at the front of the compressor the pressure they are generating suddenly reduces and there is a reversal of air flow forwards from the middle and rear of the compressor generating a large impulsive forward force onto the front stage blades. This reversed flow ceases when the driving forward pressure in the middle of the compressor has been expended. The flow then builds up again in the rearward direction until the first stages again become stalled.

The frequency of this forward/reversed flow cycle in an engine is typically 2 to 10 cycles/sec (depending upon the volume of air in the engine ducting and combustion chamber). If this were allowed to continue for, say 10 to 15 seconds, severe overheating of the air within the compressor would occur and this, together with the impulsive forces generated, would almost certainly lead to mechanical failure of the aerofoils themselves or their attachments to the discs.

The air incidence angle onto the cambered aerofoils is as good and simple an indication as anything else of the margin between stable operation of the compressor and the condition initiating surge. These incidence angles are calculated as follows :

The relative inlet angle $\alpha°_1$ is calculated from the velocity triangles (cf *Figs 6 and 11*):-

$$\mathrm{Tan}\ \alpha°_1 = \frac{U_m - V_{a_1}\ \mathrm{Tan}\ \alpha°_4}{V_{a_1}}$$

$$\mathrm{Tan}\ \alpha°_1 = \frac{U_m}{V_{a_1}} - \mathrm{Tan}\ \alpha°_4 \qquad \alpha°_4 = 25°$$

For the Rc = 15 axial compressor at 50% N_D the value of $V_{a_1}/U_m = 0.37$ (cf *Fig 10*).

$$\mathrm{Tan}\ \alpha°_1 = (1/0.37) - 0.466 = 2.237$$

$$\alpha°_1 = 66°$$

Since the rotor aerofoil inlet angle is 46° (cf *Fig 6*), the incidence angle i°

$$i° = 66° - 46° = +20°.$$

The incidence angles onto the first stage rotor and stator blading have been calculated for all the fractions of design speed given in the table *Fig 10* for the three axial compressors of design pressure ratios Rc of 3:1; 15:1; and 20:1. They are shown plotted in *Figs 12, 13, 14* respectively.

Also shown in these figures are the incidence angle i° onto the last stage rotor and stator blading. These are nearly all negative incidence angles.

The cause of these variations in positive and negative incidence angles is due to the changes in axial velocity at entry to and exit from the compressor with variation of compressor speed shown in the lower part of the *Figs 12, 13, 14* plotted against the mean blade speed Um. In all cases the middle stages of the compressors follow a straight line for the reasons previously stated (cf Chapter 3). As compressor design pressure ratio is raised the divergence of the first and last stage axial velocities from the design value Va/Um = 0.67 increases markedly as compressor speed is reduced.

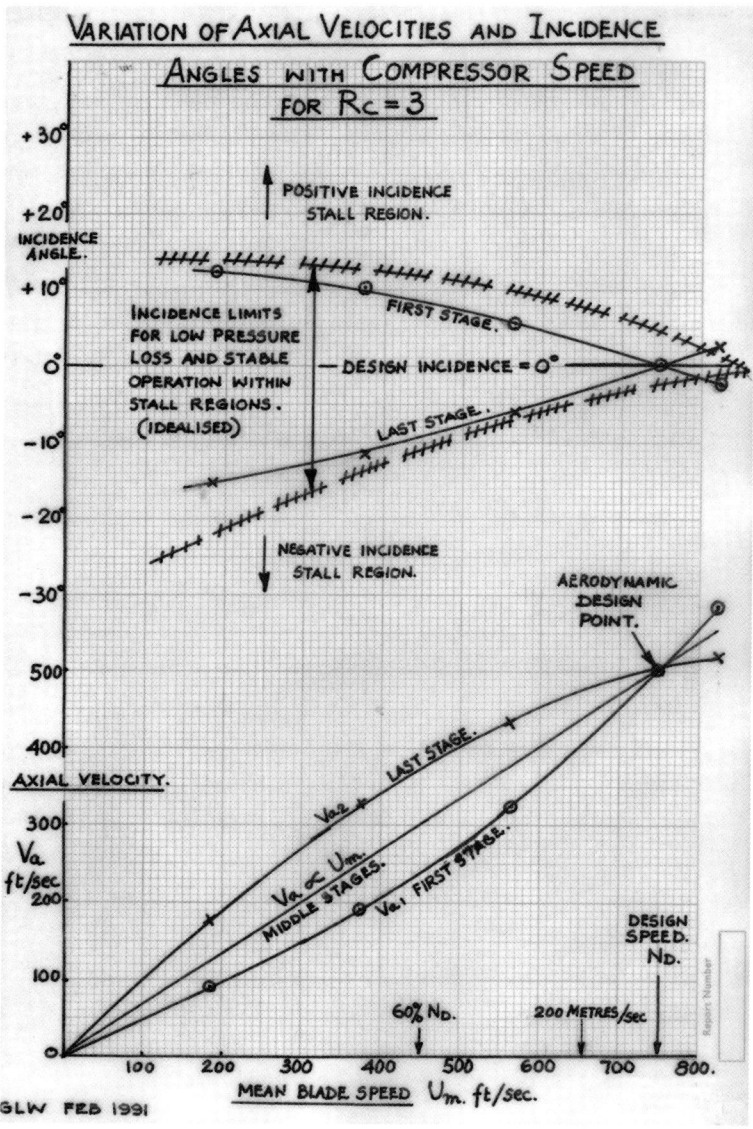

Fig 12 Variation of axial velocities and incidence angles with compressor speed for pressure ratio Rc = 3

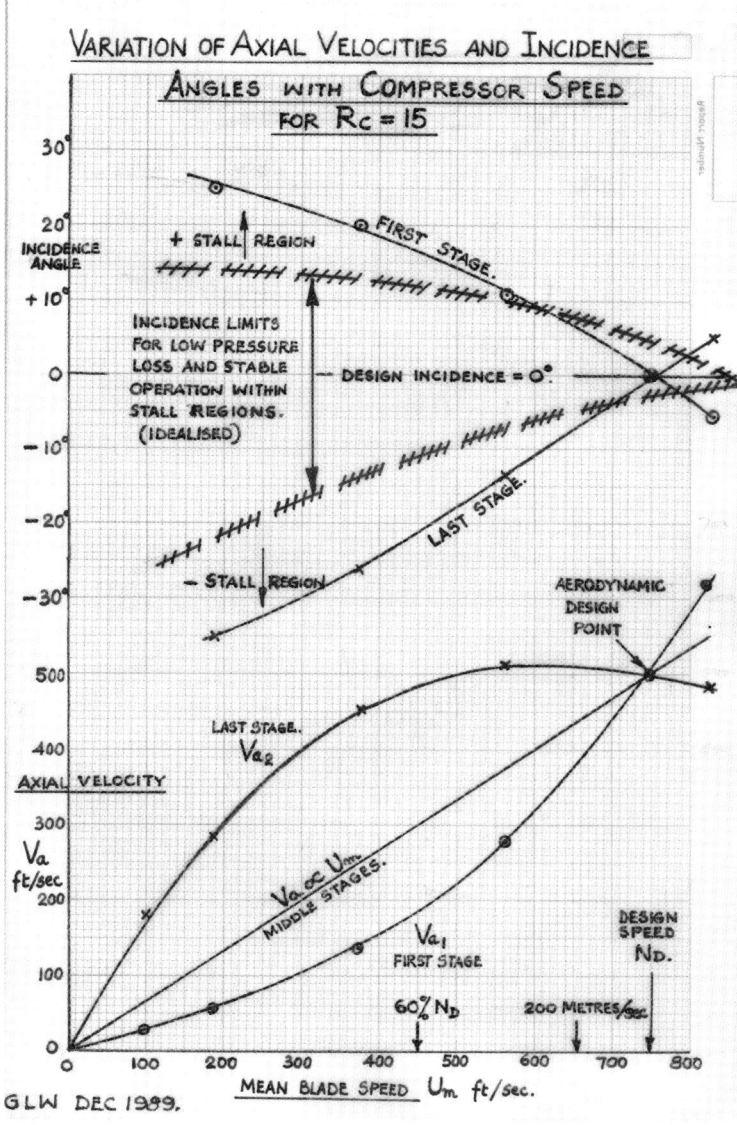

Fig 13 Variation of axial velocities and incidence angles with compressor speed for pressure ratio Rc = 15

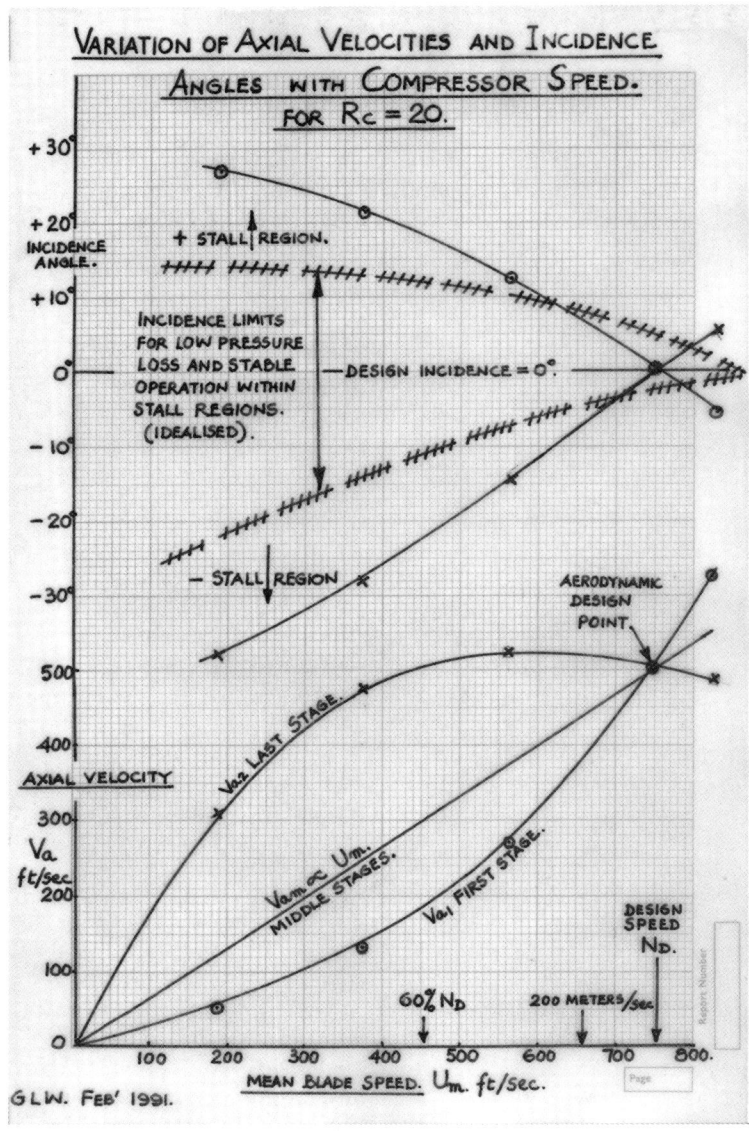

Fig 14 Variation of axial velocities and incidence angles with compressor speed for pressure ratio Rc = 20

CHAPTER FOUR

Performance of compressor aerofoils

It will be helpful to Chapter 3 to give an indication of how compressor blading performs when the air angle approaching the leading edges of the aerofoils varies from the design angle. As already indicated the difference between the air angle approaching the aerofoil ($\alpha°_1$) and the angle of the leading edge of the aerofoil (both measured relative to the compressor axis) is the incidence angle ($i°$). This is illustrated in *Fig 6*. Incidence angles can be positive or negative.

The purpose of the aerofoils is to deflect the air by as large an angle as possible without generating excessive friction and turbulence which increase total pressure losses. In order to provide test data on aerofoils for compressor design purposes wind tunnel tests are carried out by blowing air through a group or cascade of 10 or 11 aerofoils in a row enclosed in a duct. The air deflection that they impart, the total pressure loss from entry to exit, and the static pressure rise are all measured over a wide range of air incidence angles ($i°$) and inlet velocities (V_1) expressed in terms of V_1/Vs(Mach No) (Note Vs = velocity of sound in air). Thousands of tests have been carried out covering many different shapes of aerofoils and spacings to discover which performed with the least pressure losses in order to yield the highest compressor efficiencies.

Fig 15 illustrates typical test results from such tests for the pressure losses (P_1-P_2) divided by the incident dynamic pressure (P_1-p_1) = $½\rho_1V_1^2$ (for incompressible flow). As it is impossible in practice to operate a compressor within the narrow band of incidence angles necessary to ensure minimum total pressure losses at all conditions, a wider band of incidence angles has to be allowed. This is prescribed by defining the allowable total pressure loss as being twice the minimum loss. It is an arbitrary rule but one which is found to be a good practical guide for compressor design bearing in mind that an aero engine and its compressor have to operate over a wide range of conditions from ground level to altitude. The manner in which the pressure loss varies with air incidence angle and approach Mach No is shown in *Fig 15* for a particular cascade of aerofoils. The drastic reduction in incidence range with increasing approach Mach No is clearly shown.

The inset diagram in *Fig 15* plots the range of incidence angle against approach Mach No which defines an area within which the total pressure loss is equal to or less than twice the minimum loss. If the blading of a compressor could operate within this area at all the important flight conditions a high compressor efficiency would be assured. Unfortunately

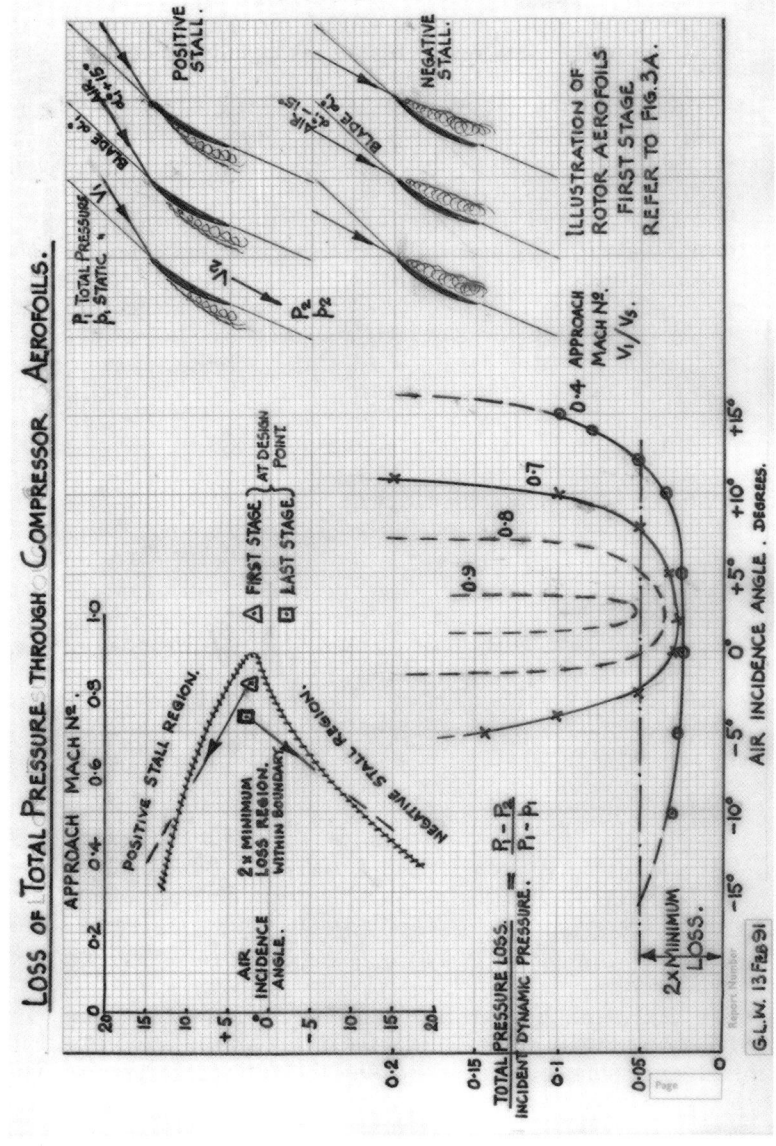

Fig 15 Loss of total pressure through compressor aerofoils. (NB for Fig 3A refer to *Fig 6*)

this is not always possible.

In order to attain the required design pressure ratio of a compressor at the design speed with the least number of stages to reduce engine weight and cost, the first stage of the compressor blading is usually placed at a position such as Δ on the diagram. As the compressor is reduced in speed, this first stage will operate from this point along the line shown in the direction of the arrow, and may be forced outside the boundary line into a region of higher incidence angles and higher total pressure loss, and even perhaps into a seriously stalled state. For the idealised compressor designs considered in this paper in which the mean axial velocity and mean blade speed are each constant throughout all the stages at the aerodynamic design condition, the last stage of the compressor would be at a point such as □ depending upon the design pressure ratio of the compressor. This is because, due to the temperature rise through the stages, the approach Mach No onto the last stage is less than that onto the first stage. As compressor speed is reduced, the last stage of the compressor will operate from this point along the line shown in the direction of the arrow towards the boundary line and possibly crossing it as indicated to the region of higher pressure loss and negative incidence stall.

It does not need much imagination to appreciate that compressor stage matching taking into account all the required engine operating conditions and the performance characteristics of the blading, (in practice along the full length of the blading from hub to tip) is a complex task of three dimensional aerodynamic flow. For this paper, it is essential to simplify the problem to a considerable extent, but without losing the fundamental stage flow mismatching characteristics that occurs when the compressor is operating well below the design speed. For this reason and to reduce the computational task, it will be assumed that point □ in *Fig 15* for the last stage coincides with point Δ for the first stage, and the cross hatched boundary line defining the air incidence angle limits for total pressure loss = 2 x minimum loss will be redrawn to a base of compressor mean blade speed instead of approach Mach No. This is an approximation that is permissible for this paper because the compressor and stage matching are being considered at ground level only, altitude conditions being disregarded.

The simplified plots of air incidence angles against mean blade speed are shown in *Figs 12, 13* and *14* for compressors with design pressure ratios of 3:1, 15:1 and 20:1 respectively. There is a dramatic difference between the diagrams for the 3:1 compressor compared with the 15:1 compressor. The calculated first and last stage air incidence angles for the 3:1 compressor are within the low pressure loss boundary over the entire speed range, whereas for the 15:1 compressor the first stage crosses the boundary at Um = 600 ft/sec or 80% of design speed and would probably be completely stalled at

60% design speed. The consequence of this would be that the compressor would surge generating violent destructive pressure oscillations. For the 20:1 compressor the conditions are only a little worse which reflects the overall density ratio of the air between 15:1 and 20:1 design pressure ratio being much less than between 3:1 and 15:1 pressure ratio (cf *Fig 9*).

Each of the *Figures 12, 13 and 14* also show how the axial velocities in the first and last stages change with compressor speed and that the axial velocity is proportional to blade speed in the middle stages, a factor supported by the analysis of compressor performance on test.

This paper is concerned only with finding a practical solution to preventing the stalling of the first stage of the compressor as the compressor speed is reduced below the design speed. The operating condition of the last stage at negative incidence is not part of this paper. It has already been stated that the assumption that points □ and Δ in *Fig 15* coincide for the calculation of *Figs 12, 13 and 14* and gives excessive values of negative incidence for the last stage relative to the negative incidence boundary lines, particularly for the 15:1 and 20:1 pressure ratio cases.

CHAPTER FIVE

The design problem

The problem is to select an axial compressor of given overall pressure ratio and air flow at the design speed N_D and then to ensure that the first stage is not in a stalled condition above 40% of N_D. This paper ignores the operating conditions of the last stage.

We have seen the velocity triangles applying to the first stage at design speed N_D and at 50% N_D, and we can now be introduced to the effect on the air incidence angles of varying the angle of the stator aerofoils and how this can reduce the incidence air angles onto the rotor aerofoils.

There is also the possibility of artificially increasing the axial velocity Va_1 through the first stage which also reduces the air incidence angles onto both the stator and rotor aerofoils. To increase Va_1 it is necessary to allow some of the air in the front half of the compressor to escape to atmosphere through ports or passages in the compressor casing.

The possibilities are illustrated in *Fig 16* and it should be noted that a combination of varying the angle setting of the stator aerofoils together with allowing a proportion of the air to escape to atmosphere (usually referred to as 'bleed air') is a further possibility that has been used in practice with success.

To simplify the problem, the air flows involved, the velocity triangles, and forces are all concentrated at the mean radius Rm otherwise the problem would become unmanageable for the purposes of this paper.

The problem also includes an assessment of the forces applied to the stator aerofoils and how the couple and bending moment created are to be resisted by the stator blade swivelling fixing in the compressor casing.

If bleed air is used the bleed air ports in the casings and the design of a suitable air bleed valve should be proposed that is open up to 40% N_D and closed above this speed. *Fig 17* illustrates these aspects.

It is suggested that the following data is given :

1. Inlet pressure (P_1) 14.7 lb/sq.in absolute.
2. Inlet air temperature (T_1) 288°K and $\gamma = 1.4$ constant.
3. Air mass flow (M) at design point 100 lb/sec.
4. Design mean blade speed (Um) 750 ft/sec.
5. Design axial velocity (Va) in all stages 500 ft/sec.
6. Overall design pressure ratio (Rc) 12:1.
7. Design temperature rise per stage in the first half of the compressor 20°K.

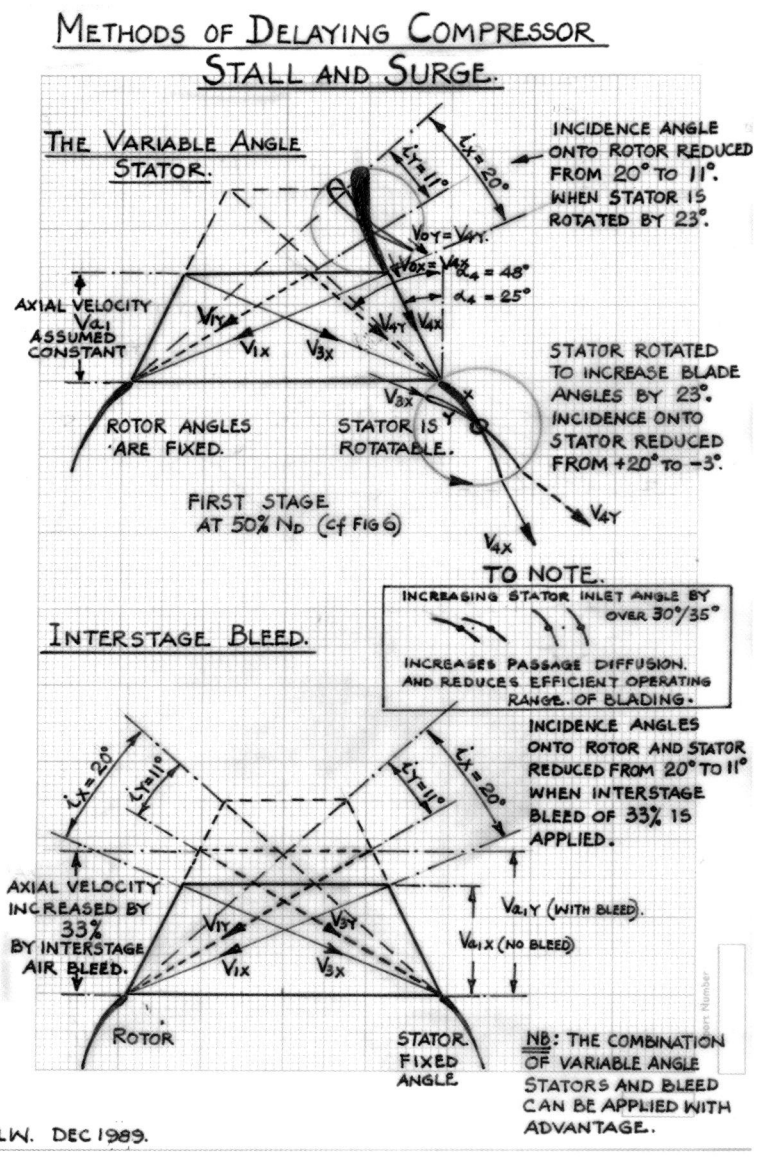

Fig 16 Methods of delaying compressor stall and surge

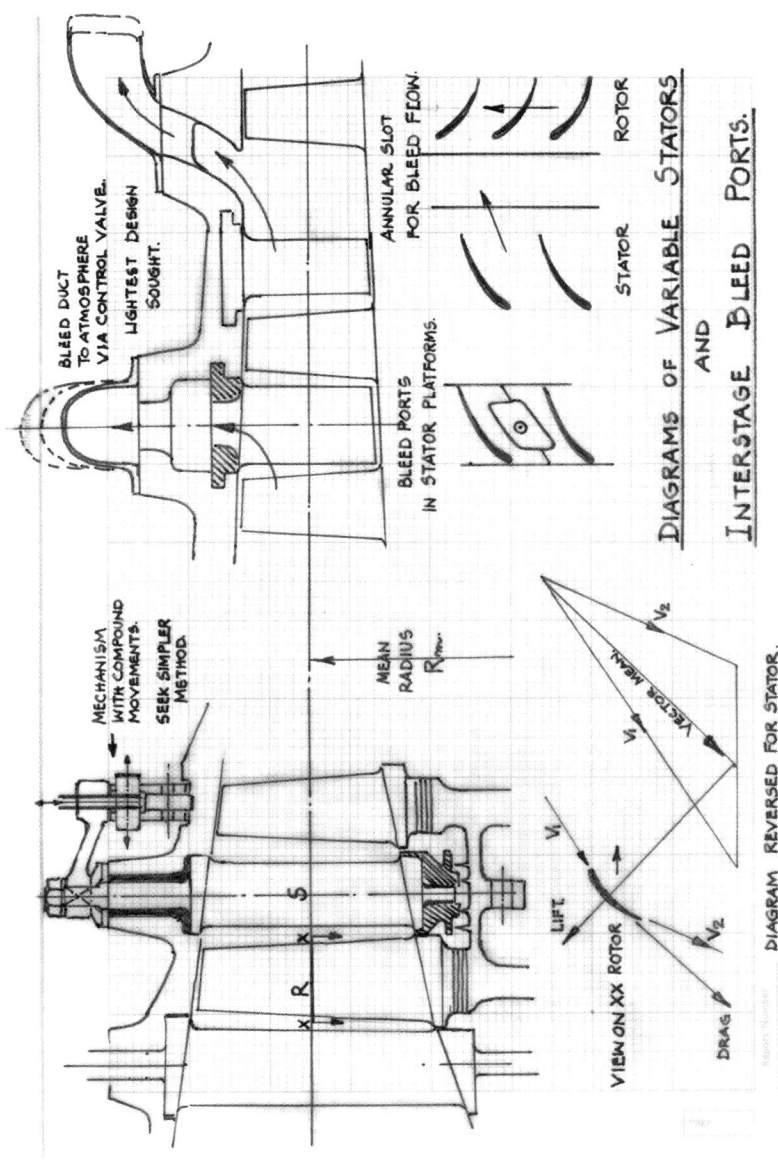

Fig 17 Diagrams of variable stators and interstage bleed ports

8. Number of compressor stages 19. This requires the temperature rise per stage in the last three stages to be reduced slightly to give an integral number of stages.
9. Compressor bleed air, if used, to be taken from stage 6 where the temperature has reached $T_1 + 6 \times 20°K = 288 + 120 = 408°K$ at design speed, but less at lower speeds.
10. Stage adiabatic efficiency $\eta_{ad} = 85.72\%$ treated the same as polytropic efficiency η_p. Hence the relationships between pressure ratio, density ratio etc., given in Chapter 3 apply.
11. The compressor outlet hub/tip ratio $Rh/Rt = 0.80 = Y$ from which the constant mean radius Rm can be found from $M = A_2.Va_2 \rho_2$ and $A_2 = \pi Rm^2 (1/Y - Y)$. Rh and Rt can now be found for the first stage which establishes the radial length of the first stage rotor and stator blading (see page 35). The length of any other stage along the compressor could be found in a similar way.
12. The mean chord of the first stage rotor and stator blading should each be 1/3 of the annulus radial length at the centre of the stage.
13. The stator blades have a circular arc camber line and must have a maximum thickness to chord ratio not exceeding 15%. The material is titanium. A maximum working tensile stress of 20 tons/sq.in (309 MPa) can be assumed.
14. The force acting upon the stator due to the deflection of the air (*Fig 17*) is assumed in this paper to be concentrated at the mean radius Rm. This force is then doubled to allow for the impulsive force applied forwards when the flow direction reverses due to aerofoil stall and the subsequent surge pulse. This is an empirical factor from experience.
15. The rotor and stator blading design in this exercise conform to 50% reaction design which makes the static pressure rise across rotor and stator equal. This requires that there is an inlet guide vane immediately ahead of the first rotor to direct the air at a whirl angle of $\alpha°_o$ equal to the air angle $\alpha°_4$ leaving the first stator to approach the second rotor (cf *Fig 16 TOP*).

In this paper, in order to limit the computational task, only stage 1 has been considered. However, it should be understood by the engineering student that at compressor speeds below design, the flow mismatch of the stages between stage 1 and the middle stages diminishes progressively, and that this also applies between the last stage and the middle stages.

In practice therefore, if variable angle stator blading is adopted to delay the onset of stall as compressor speed is reduced below design speed both the inlet guide vanes ahead of the first rotor and the first stator are variable and move together. The succeeding three or four stage stator (stages Nos. 2, 3, 4,

5) must also be variable, but the angle through which they need to be rotated progressively diminishes from stage 2 to stage 5.

It will be obvious that the number of variable stators required is a function of the design pressure ratio Rc of the compressor. In practice the working out of such designs stage by stage is a vast computational task. It requires the three dimensional flow in the annular flow passage of the compressor covering the full radial length of the blading to be taken into account.

As the design pressure ratio of the compressor is raised, above say 12:1, it is advantageous to apply both variable stators and interstage bleed to the compressor. This is because the angle through which the stators are turned, (and particularly those in the first three stages) can be limited to angles at which the stators perform efficiently. Reference to *Fig 16* will help to explain this.

It should now be possible for a third year degree student with his background of academic studies in thermodynamics, fluid mechanics and general mechanics to work through the case suggested and offer design solutions. The sketches in *Fig 17* are intended as a guide.

Clarification of hub/tip ratio

The hub/tip radius ratio of an axial compressor stage is a significant factor for the following reasons:-

(a) Above a value of $Rh/Rt \doteq Y$ of 0.8 the air flow between the blades at the outer and inner ends is irregular due to boundary layer development causing reduction in stage efficiency particularly in the high pressure stages of a high pressure ratio axial compressor.

(b) High duty compressors for aero jet engines aim to deliver the highest possible air flow rate within engine diameter limitations. The hub/tip of the first stage may be as low as 0.3. This complicates the aerodynamic design of the blading which must now take account of three dimensional vortex flow factors. The resulting blading is highly twisted radially and compressibility effects require the aerofoil sections to change considerably from hub to tip.

(c) The mechanical design of the lower hub/tip ratio stages is complex due to high stresses, aerodynamic flutter and mechanical vibration.

For this simplified exercise, the hub/tip ratio factor Y is introduced as follows:-

The required annulus areas through the stages are calculated from air temperature rise at the stage, the assumed efficiency and the axial velocity:-

$$\text{Annulus area } A = \pi (R_t^2 - R_h^2)$$

$$= \pi R_t^2 \left(1 - \frac{(R_h)^2}{(R_t)^2}\right)$$

$$\text{Mean radius of annulus} = R_m^2 = R_t \cdot R_h = R_t^2 \cdot \frac{R_h}{R_t}$$

$$\therefore R_t^2 = \frac{R_m^2}{R_h/R_t}$$

$$\therefore \text{Annulus area } A = \pi R_m^2 \left(\frac{1}{R_h/R_t} - R_h/R_t\right)$$

$$A = \pi R_m^2 \frac{(1 - Y)}{Y}$$

Reference to *Fig 9* illustrates the relationship between annulus area, hub/tip ratio of stages for five axial compressors with the same mean radius R_m and overall pressure ratios R_c of:- 3; 6; 10; 15; 20:1 respectively. Also they all have the same outlet annulus area A_2.

CHAPTER SIX

The pressure ratio – flow characteristics of axial compressors

The pressure ratio-flow characteristics of axial compressors (usually just called 'characteristics') are obtained by driving the compressor at given rotational speeds in steps over a wide range of speeds (typically 40% to 110% of design speed) on a special test plant. At each speed and at a given selected intake pressure (P_1) the delivery pressure (P_2) is varied by throttling the delivery flow progressively from the fully open area setting to a minimum area setting. This minimum area setting of the delivery throttle is found when the compressor flow and delivery pressure become unstable, often with violent pressure oscillations. The pressure ratio and flow at which this occurs is recorded. This must be followed by rapid opening of the throttle to move from this condition to avoid damage to the blading. Stable conditions are then restored although sometimes there is a marked hysteresis effect with stable conditions occurring at a marked lower delivery pressure. The pressure ratio and flow at which instability occurs is called the 'surge point'. This has to be recorded at each speed tested because the locus of the surge points (called the 'surge line') over the speed range is vital information for determining whether the necessary engine design pressure ratio and flows required to attain the design performance of the engine can be realised in practice.

Fig 18 is a diagram of the Rolls-Royce compressor test rig that was built at Sinfin, Derby, to test the new AJ65 compressor. It was powered by BTH 5000hp electric motors and known as No 3 rig. It supplemented the existing No 2 rig at Sinfin, which was powered by English Electric 2000hp electric motors. At the design stage of the AJ65 engine, the Compressor Development Group in Derby had planned to rig-test the compressor shortly before the engine ran at Barnoldswick. Wisely, they had planned to test the compressor in three separate parts – the first four stages on their own (on No 3 rig), the last eight stages on their own (on No 2 rig) and the complete 12 stage unit (on No 3 rig) with the object of measuring the separate pressure ratio-flow characteristics and efficiencies and the flow-matching of the parts. This proved most effective because they were able not only to warn Barnoldswick in advance that the engine would be in surge trouble, but they also knew the reason.

The flow range of the first four stages was restricted due to premature stall of the blading and when this occurred there was a sharp reduction of pressure causing the surge line of the 12 stage unit to drop below the compressor working line.

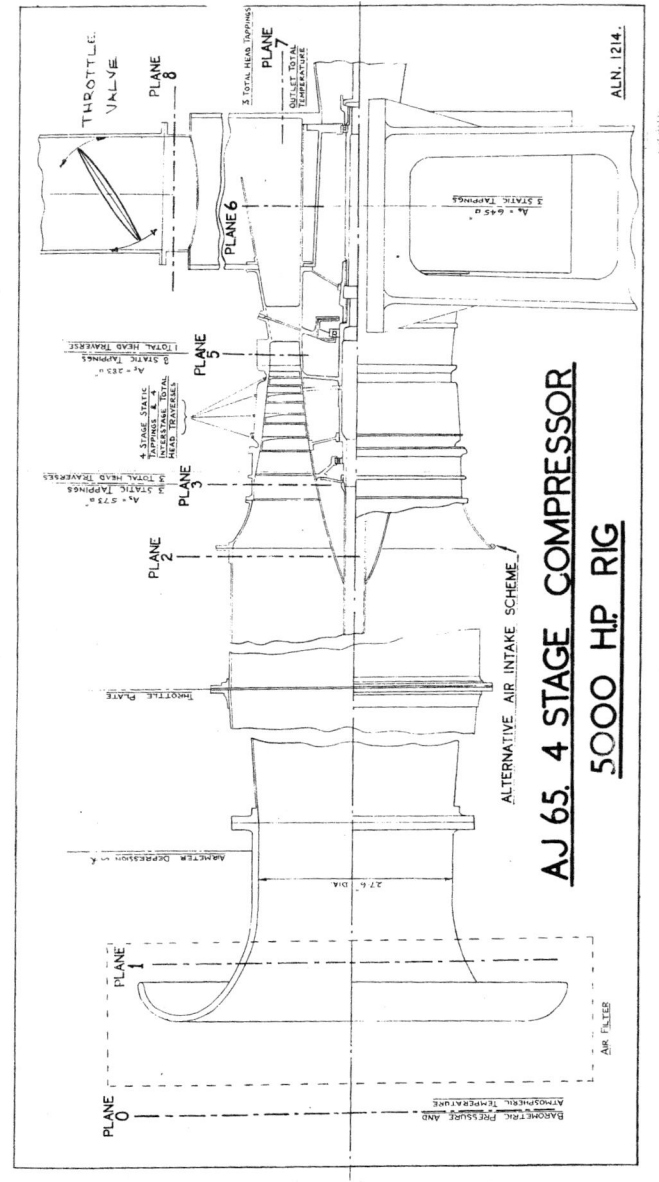

Fig 18 The Rolls-Royce 5000hp compressor test rig

A number of proposals were made for different blade designs to increase the flow range and postpone stall. These designs were made in a very short time for rig test, but none were successful. A series of tests were then initiated in which the first five stages of stator blade rows were set at progressively increasing stagger angles from stage five to stage one. A number of rows of stator blades set at angles of +5° to +30° in 5° steps were made and a series of 12 stage tests were carried out to find the best combination for surge-free operation of the engine. From these tests, it was shown that if the compressor was designed with controlled variable angle stator blades, the surging of the compressor on the engine could be avoided. As we shall see in the next chapter, they went on to build an experimental variable stator compressor to prove the point.

However, during this rig work, they had also shown that the AJ65 compressor with a variable inlet guide vane only combined with interstage bleed would give surge-free operation on the engine and this method was adopted because it was a simpler solution for this engine with a pressure ratio of 6.3:1.

The measurements taken during this series of rig tests (and other rig tests at Rolls-Royce) are:

(a) Inlet air total pressure and temperature.
(b) Rotational speed.
(c) Inlet air mass flow. Air is filtered before entering the venturi air meter.
(d) Outlet air total pressure and temperature. The temperature is measured inside 'stagnation vorticity' pockets.
(e) Input shaft power by means of optical or electrical torquemeter.
(f) Interstage static pressure and total temperature.

From these measurement two adiabatic efficiencies are calculated, one based upon the temperature rise and the other based upon input shaft power. One is a check upon the other. It is vital to obtain the most accurate assessments of efficiency to chart development trends. The efficiency based on shaft power may be 2 to 3% lower than that based upon temperature rise partly explained by windage and mechanical losses. When overall efficiencies are in the region of 88 to 90% individual design improvements are not likely to exceed 1 to 2%. Therefore it is essential to strive to measure efficiency to an accuracy of $\pm\frac{1}{2}\%$ otherwise it becomes impossible to decide whether a design modification is an improvement or not. If there is inconsistency of results it is sometimes necessary to dismantle the compressor, rebuild it, and repeat the tests. This is expensive in time and effort.

The interstage static pressures and temperatures are taken so that estimates can be made of the individual stage characteristics. These will indicate the relative flow matching of the stages and the flow at which individual stages stall. This can sometimes explain why the compressor surges abnormally at a particular speed and provide the data to decide what change in stator or rotor blade angle should be made to delay the stall.

Fig 19 is a typical plot of a set of compressor characteristics for a 7 stage compressor designed to produce a working pressure ratio in the engine of 5:1. This is a compressor with quite a high aerodynamic duty (high pressure ratio per stage). As speed is reduced below design speed the surge line, shown by the full line, dips to lower surge pressure ratios below 87% speed, and at 77% speed has dropped to intersect the steady state compressor working line which is determined by the turbine nozzle flow characteristic. At this condition the front stages of the compressor are stalling causing a drop in compressor efficiency and initiating compressor surge. The compressor delivery pressure and engine become unstable with flow and pressure oscillations developing. There is the danger of damaging the compressor and overheating the turbine caused by the rise in combustion outlet temperature with oscillating flow. The fuel supply to the combustion chamber must be cut and the engine shut down.

With the surge line coinciding with or falling below the steady state compressor working line it is not possible, after starting the engine by means of a powered motor, to accelerate the engine under its own power. This is because additional fuel over and above that required for steady state running is required to provide the additional turbine driving torque for acceleration. When the compressor is stalled the injection of extra fuel into the combustion chamber only forces the compressor into a more deeply stalled state causing overheating of the turbine.

With the surge line placed well above the steady state compressor working line, which is the required normal state at all speeds, injecting additional fuel for engine acceleration moves the working point to higher pressure ratio on the compressor characteristic as shown at A in *Fig 19*. In this example the object is to increase the engine speed from A to B. As B is approached the excess fuel is trimmed back to the steady state fuel flow rate. The time interval between A and B may be only 1 second. The injection of the additional fuel for acceleration must not be so large as to raise the transient pressure ratio on the characteristic up to the surge pressure ratio. With a manually controlled engine throttle there is the danger of this occurring. To overcome this risk the engine fuel control systems now usually incorporate an 'acceleration fuel limiter' which prevents the pressure ratio increasing above the indicated dotted line above AB in *Fig 19*.

It should be clear from the foregoing that the stalling of compressor blading must be prevented below 87% design speed to avoid the onset of

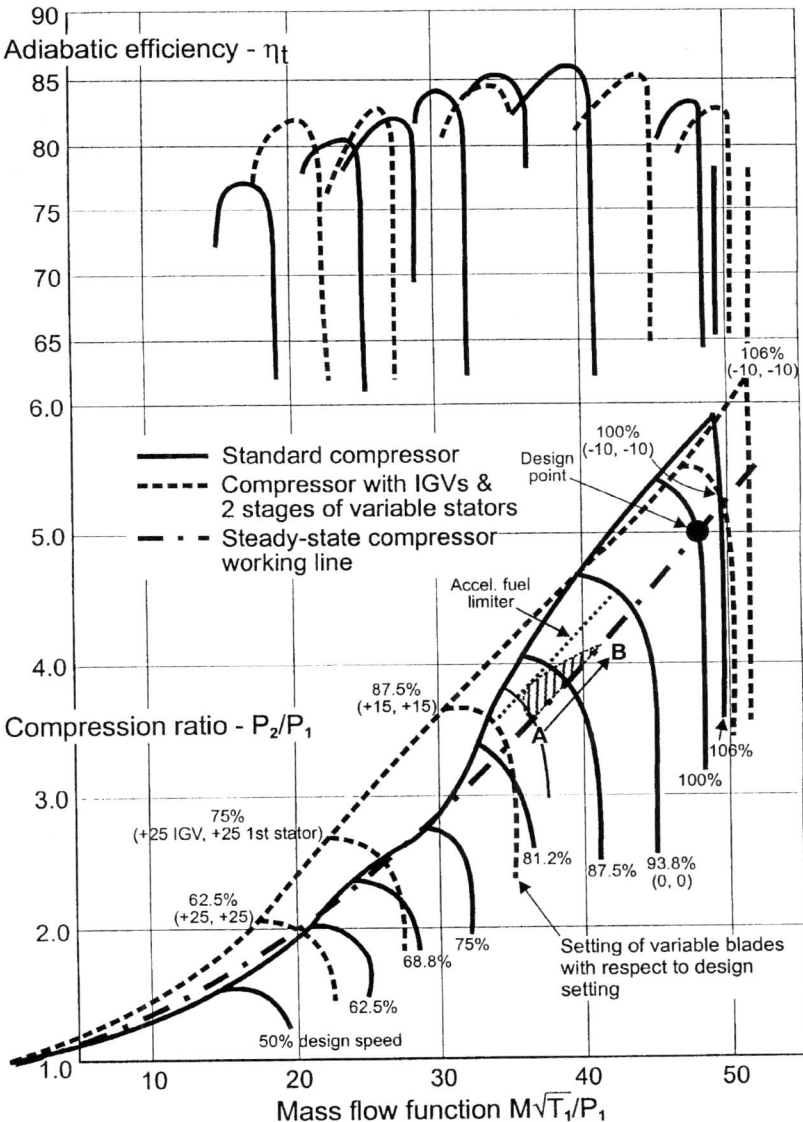

Fig 19 The performance characteristics for a seven-stage 5.0:1 working pressure ratio compressor

compressor surge. This is done in this case by designing the compressor with a variable angle inlet guide vane followed by variable angle stator blade rows in stages 1 and 2. These movable blade rows are regulated by a servo control system which moves them progressively with engine speed (rpm corrected) from the design angle (+0°) at 87% design speed to (+25° higher than design) at 75% design speed and below. The effect is to raise the surge line to the dotted line in *Fig 19* providing ample excess pressure ratio above the steady state compressor working line to absorb the additional transient 'acceleration fuel' without surge.

Once variable angle stator blading has been introduced, the vanes can be used beneficially to a limited extent to increase the mass flow and pressure ratio at and above the design speed to increase engine thrust. This is obtained by setting the blades at up to (-10° lower than design) as indicated in *Fig 19*.

It is hoped that this account of the operating requirements of axial compressors in an aero gas turbine engine will provide useful background to the importance of the flow matching of the stages in this paper which, albeit considerably simplified, does still bring out the fundamentals of the problem. It must also be understood that there is not any reliable theory that can predict the surge line of a new design of axial compressor. Prediction of the critically important surge line is largely empirical based upon the correlation of hundreds, even thousands of compressor rig tests. Even today after 50 years of axial compressor development serious errors can still be made in predicting the surge line of a new compressor. This is in contrast to the aerodynamic design of a compressor at a given chosen design condition which, with the current methods of computational aerodynamics and the computer can be accomplished with high precision. But the off-design performance, even using these methods, can be wide of the mark. The students should be aware of what advanced theory can accomplish, and what, through the complexity of three dimensional flow at off-design conditions, even advanced theories cannot explain. The importance of high quality experimental work and test analysis are still vital to progress. It is important to continue emphasising this because we see a tendency everywhere to spend more on developing theory and computer programmes for the design point without comparable effort being given to the more complex and less predictable off-design flow conditions.

Reference has been made to taking measurements of the interstage pressures and temperatures when carrying out compressor rig tests to obtain the overall characteristics. From these interstage measurements the individual stage pressure ratio-flow characteristics can be calculated (see Page 44 and *Fig 20*). They show the range of axial velocities (and also the air incidence angles) over which each stage is forced to operate from low rotational speeds to design speed and above, and vice versa.

To correlate the measurements at the different speeds the stage pressure ratio (Rs) is expressed in the form

$$\psi = Kp\, T_1\, (Rs^{(\gamma-1)/\gamma \cdot 1/\eta_p} - 1)/Um^2 \text{ plotted against } \phi = Va_m/Um$$

where Kp = specific heat at constant pressure (Cp) in mechanical work units.

Unfortunately it is not possible from interstage pressure measurements alone to deduce the separate stage adiabatic efficiencies, and even if temperature measurements are made by special small thermocouples placed in stagnation pockets within the stator blade passages, the results give wrong values of stage efficiencies. This is because such point readings do not give the average temperature across the flow field between the blades. Nevertheless the value of ψ, and particularly the slope of the lines through the points, do reflect the variation in total pressure losses that occur in cascades of aerofoils as ϕ, and the corresponding incidence angles on to the aerofoils, vary.

The value of ψ for each test point is plotted as ordinate against $\phi = Va_m/Um$ as abscissa. At each compressor test speed the variations of ψ and ϕ are small, but when all the points at all the speeds are put together, an approximation to the true stage characteristic can be drawn. This has been found helpful in presenting a reasonably clear picture of the flow matching of succeeding stages. The values of ϕ at which individual stages approach peak pressure and, in some cases (notably the first several stages of the compressor) follow on to be forced to operate heavily stalled, are clearly shown. This is signified by an abrupt change in slope of the characteristic from negative (stable operation) to positive (unstable operation).

The *Figs 21 to 26* show values of ψ plotted against ϕ obtained from Avon RA3 engine (cf *Fig 29*) axial compressor tests over the speed range 4000 to 7720 rpm. This compressor had a design pressure ratio of 6.3:1 and, as shown earlier in this paper, the first stage would be expected to be operating stalled below about 75% of design speed. This is shown to be the case between 6000 rpm (cf *Fig 23*) and 5000 rpm (cf *Fig 22*). On each of the *Figs 21 to 26*, the derived first stage characteristic has been drawn so that the measurement of the operating points can be traced as the compressor speed changes. At 4000 rpm the first stage is in deep stall and the last stage is approaching the condition of zero or negative pressure rise (turbining).

In *Fig 27* all the characteristic curves drawn from all the test points at all the speeds are assembled together. From this it is clearly seen that the front stages (1 to 5) inclusive are driven into deep stall as a result of being forced to operate over a wide range of ϕ (and corresponding incidence range). The middle stages (6 and 7) operate over a limited range of ϕ on either side of the design value $\phi = 0.67$ over which these stages are producing their peak pressure. The rear stages (9 to 12) inclusive are forced to operate at high values of ϕ at the low compressor speeds leading to negative incidence angles and ultimately choking (Mach No $\simeq$ 1.0 in the throat between the blades).

These experimental results are in general agreement with the simplified theory of the flow mismatching of stages of axial compressors given in this paper.

Derivation of individual stage characteristics

Early in the performance testing of the Avon 12-stage compressor on the No 3 (5000hp) compressor test plant at Sinfin, we attempted to derive the individual stage characteristics to try to explain the unexpected overall characteristics.

This was done by measuring the static pressure at the outer wall of the annulus at each stage immediately downstream of the stator trailing edges. With this pressure and knowing the air flow rate, flow area and approximate temperature, the total pressure could be calculated.

The static pressures were measured at every compressor test point at every speed. The total pressures calculated and corresponding stage total pressure ratios were used to calculate stage temperature rise ratios assuming stage efficiency consistent with the compressor overall efficiency. This, of course, was subject to error but it was convenient for correlation with compressor speed and air flow rate in trying to obtain an understanding of stage flow matching and hopefully to show when blading became stalled causing compressor flow instability and surge. In this, such as it is, the analysis proved useful (refer to *Figs No 21 to 27*). The change in slope of the characteristics of the first five stages at the lower compressor speeds is clearly brought out indicating stall leading to unstable flow. The background to the method is shown in *Fig 20*.

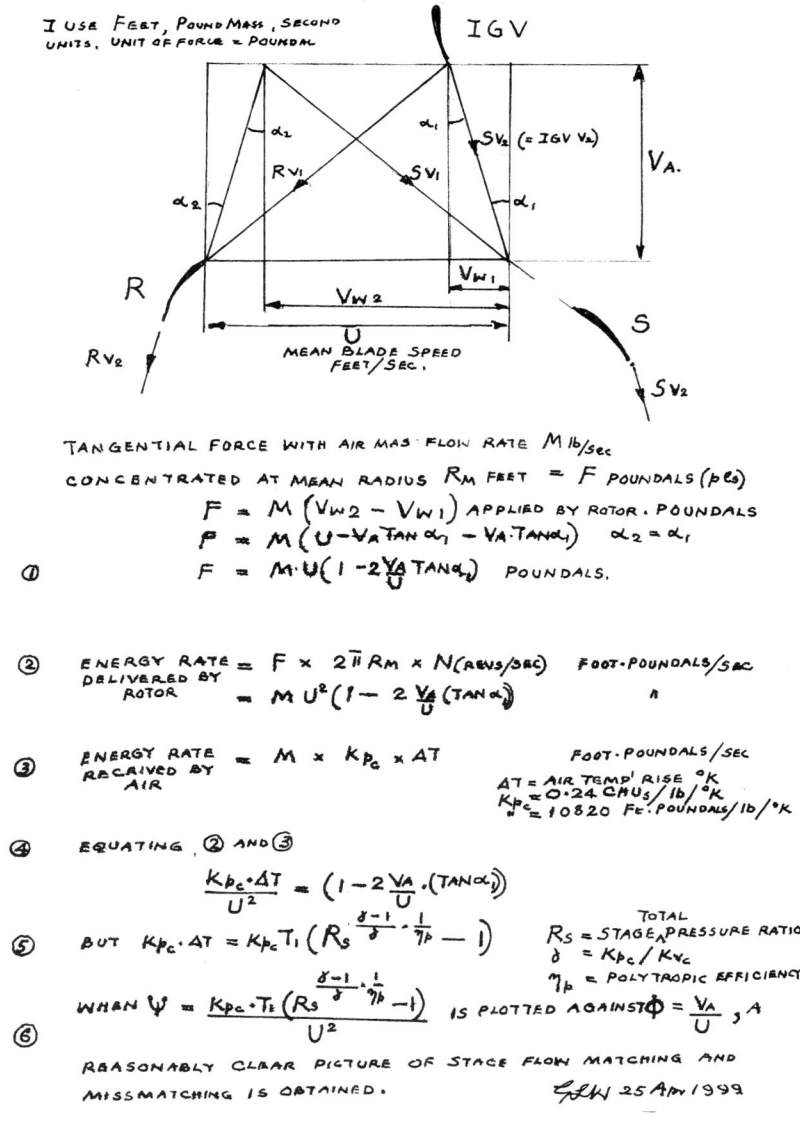

Fig 20 Derivation of stage pressure ratio function.

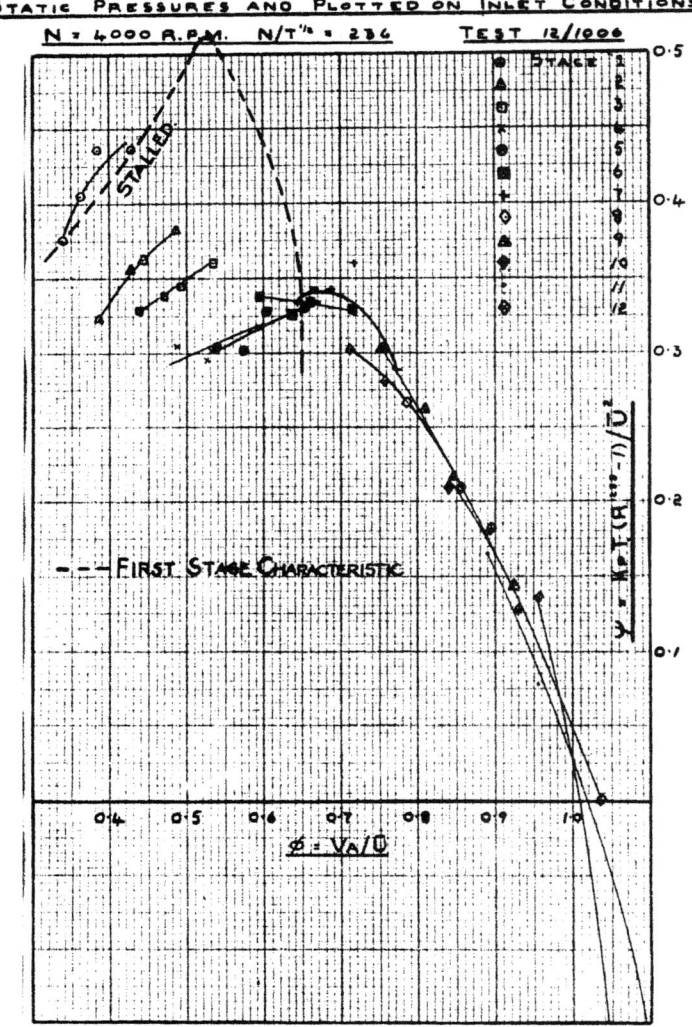

Fig 21 Rolls-Royce Avon compressor stage characteristics - 4000rpm

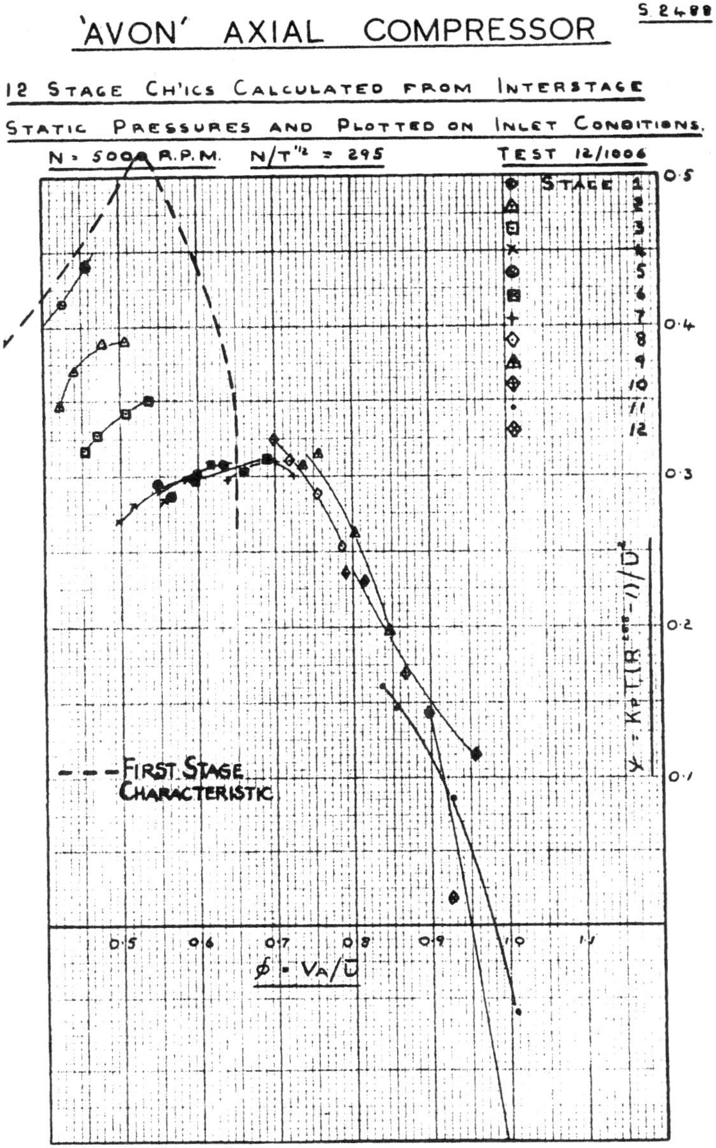

Fig 22 Rolls-Royce Avon compressor stage characteristics - 5000rpm

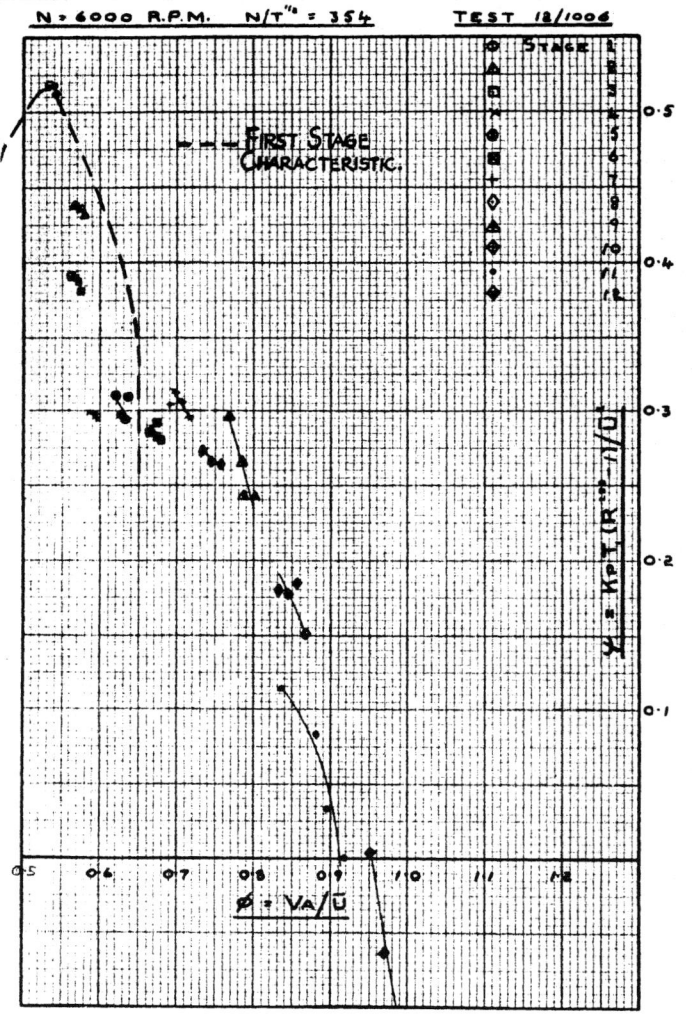

Fig 23 Rolls-Royce Avon compressor stage characteristics - 6000rpm

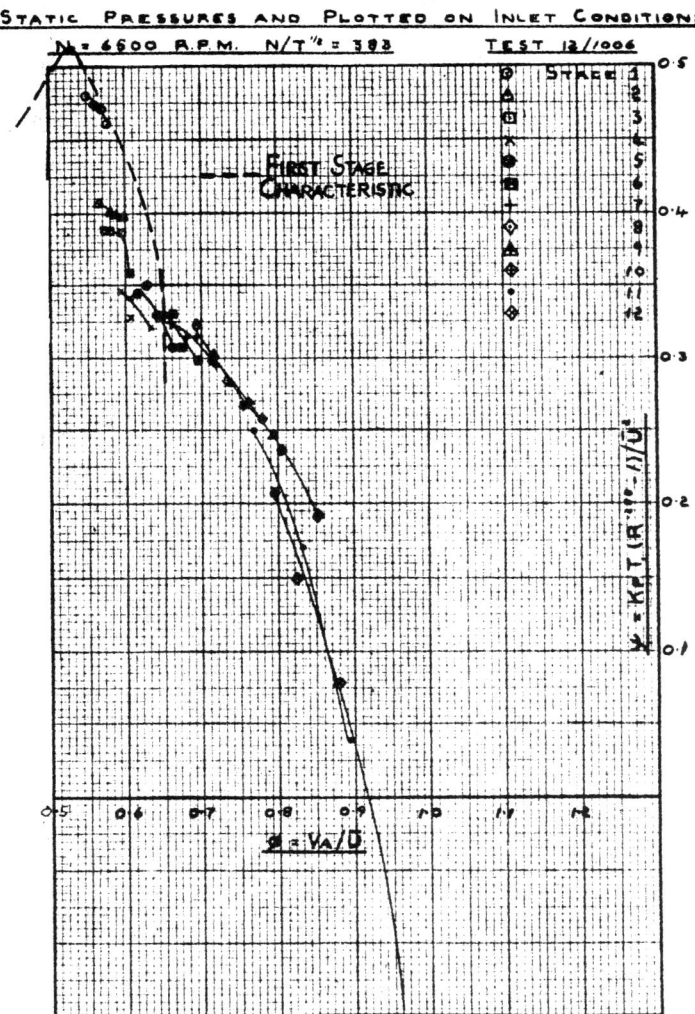

Fig 24 Rolls-Royce Avon compressor stage characteristics - 6500rpm

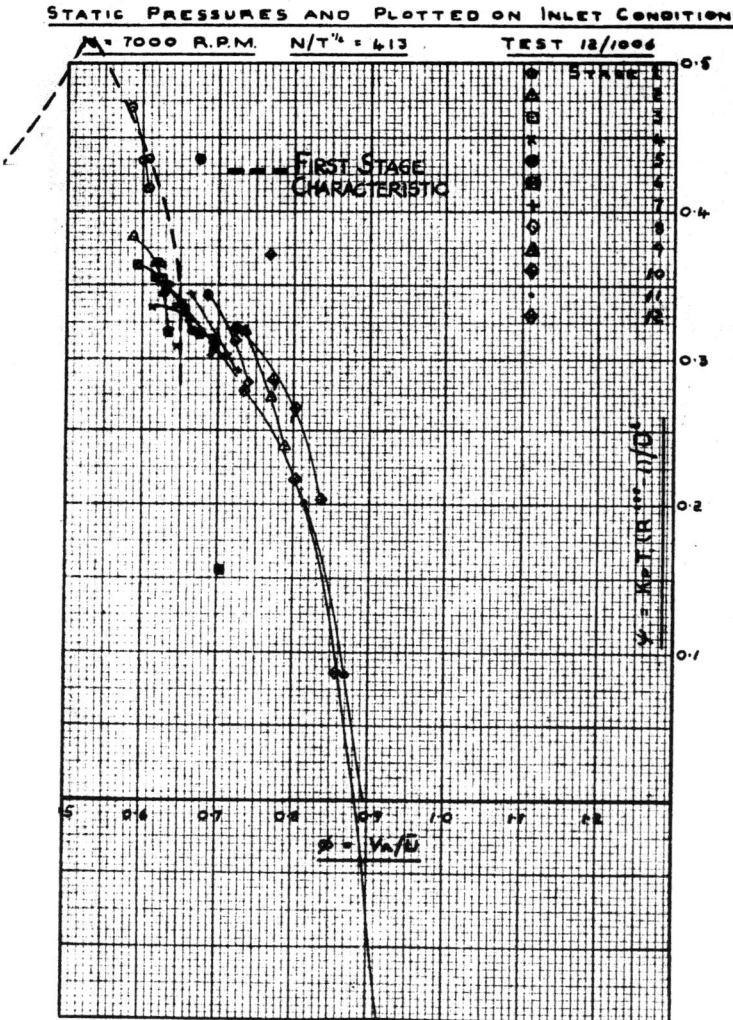

Fig 25 Rolls-Royce Avon compressor stage characteristics - 7000rpm

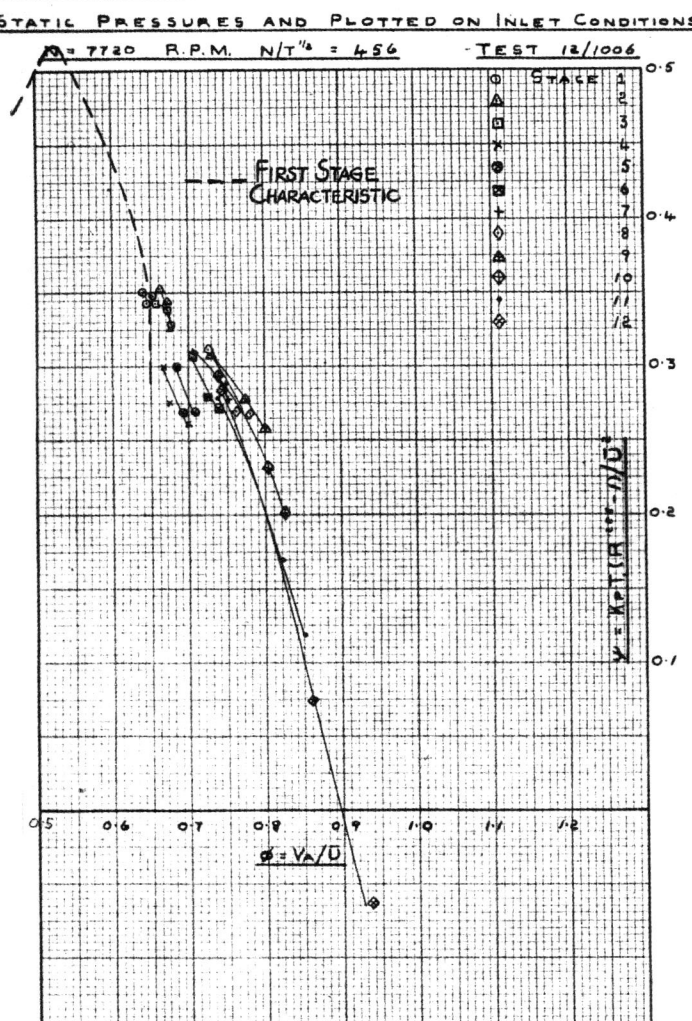

Fig 26 Rolls-Royce Avon compressor stage characteristics - 7720rpm

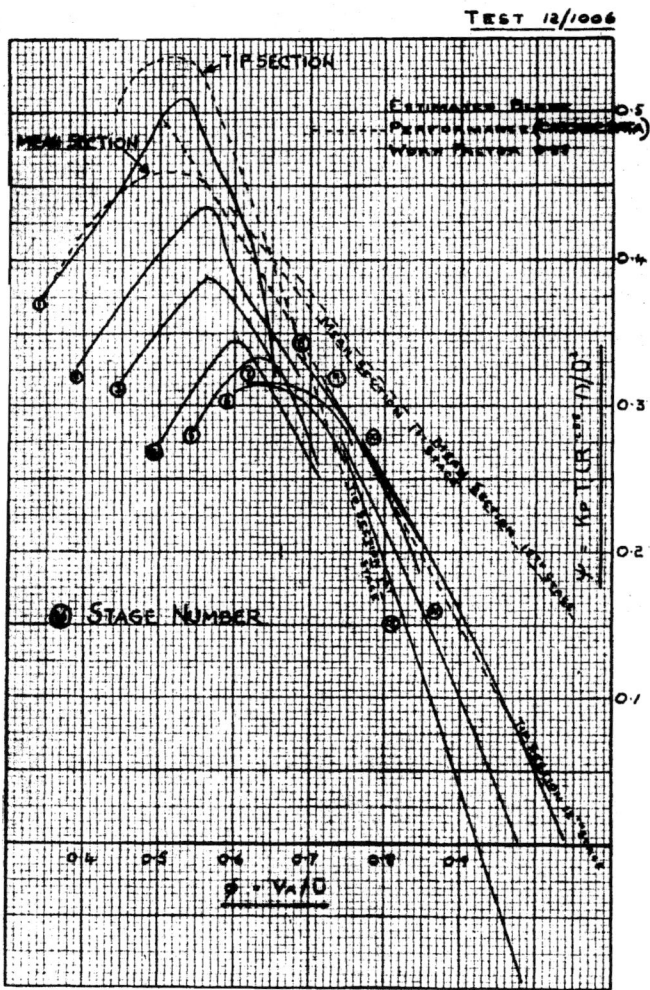

Fig 27 Rolls-Royce Avon compressor mean stage characteristics for all speeds

CHAPTER SEVEN

Examples of axial compressors in practice

The following brief descriptions of several engines with axial compressors are included in this paper. They serve to emphasise the fundamental importance of the subject of this design exercise, the flow matching of the stages at off-design conditions and how this effects the design of engines as a whole.

The first Rolls-Royce high pressure ratio axial turbojet, AJ65 (*Fig 28*) was designed in 1945. The pressure ratio was 6.3:1 (then considered high). It was an ambitious project. Apart from the advanced design of the compressor for that period, the engine had a single stage turbine of high aerodynamic loading to drive the compressor. When built and put to test the engine was found to be seriously flawed in two respects:

1. The compressor front stages were in deep stall at low engine speeds causing the compressor to surge. After starting the engine, it could not be made to accelerate to the higher speeds at which the blading would have become unstalled. The high speed performance of the engine could not be approached and assessed.

2. The single stage turbine had a low adiabatic efficiency (76%) which caused the engine to operate at higher turbine entry temperatures than expected. This raised the pressure ratio of the steady state compressor working line which added to the problem 1 above.

It was a critical situation from the engine development standpoint. Progress could not be made without major redesigns of both the compressor and turbine.

The deep stall problems of compressor blading and the sensitivity of the compressor to surge were new problems at that time. Intensive programmes of compressor and turbine rig testing on special test plants were carried out to diagnose the problems. From this work major redesigns of the compressor and turbine were carried out leading to the Avon RA3 (*Fig 29*) which succeeded the AJ65. They were:

3. The compressor was rebladed to increase the low pressure loss incidence range of the blading, and a variable angle inlet guide vane and interstage bleed were incorporated. These were controlled as a function of engine corrected speed and the deep stall of the blading and compressor surge

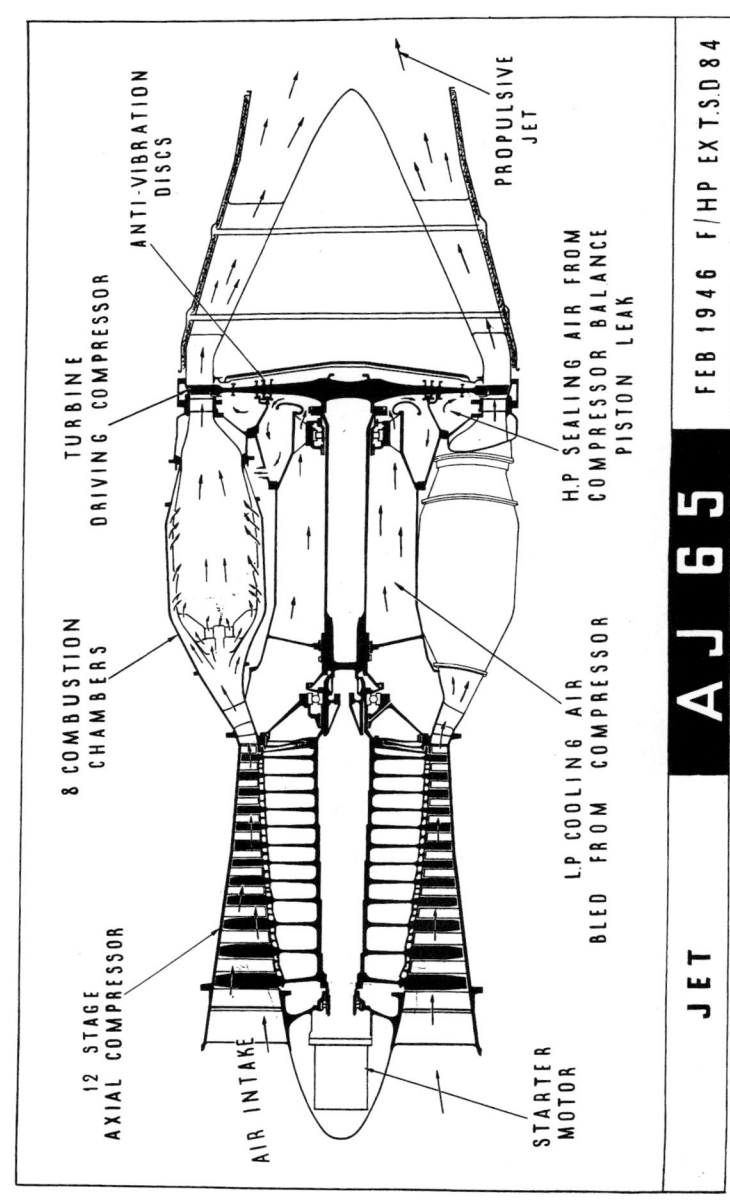

Fig 28 Cross-section of Rolls-Royce AJ65 axial turbojet

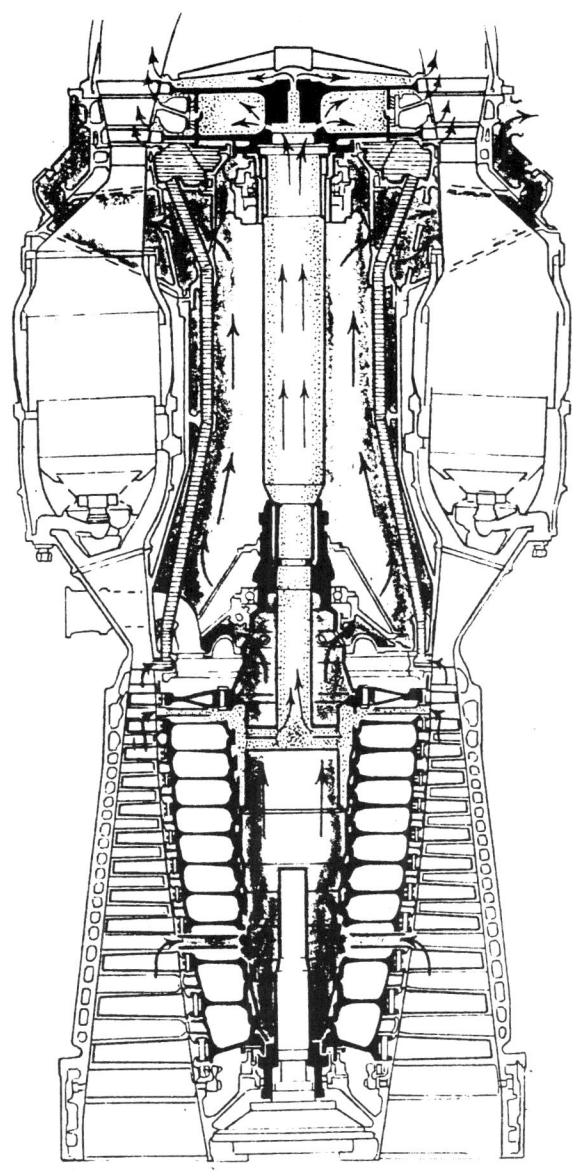

Fig 29 Cross-section of Rolls-Royce RA3 axial turbojet

were alleviated to the extent that the engine could now function satisfactorily over the required engine speed range. The design pressure ratio of 6.3:1 of the AJ65 was retained in the Avon RA3.

4. The single stage turbine was replaced by a two-stage turbine of appreciably higher adiabatic efficiency (85%). This also contributed to overcoming the low speed compressor stall problems by enabling the engine to operate at a lower turbine entry temperature and with a lower compressor working line.

The modified engine, retaining the original size and pressure ratio of the 12-stage compressor but incorporating variable inlet guide vanes and interstage bleed, went into production for the English Electric Canberra B2 (in 1950), Hawker Hunter Mk 1 (in 1951), de Havilland Comet 2 (in 1952), Swedish Lansen, Australian Sabre F86 and proved to be a successful engine.

Previous reference has been made to the intensive programme of compressor rig testing. Once the stall problems had been identified a new design of compressor was launched in which the inlet guide vanes and first four rows of stator blades could be swivelled to reduce the air incidence angles on to the blading at the lower engine speeds. This proved most effective. It was the first time in the history of axial compressor technology worldwide that variable angle stator blading had been applied to overcome stall and surge (this was in 1949). It was a major breakthrough. The unique 12 stage rig compressor is shown in *Fig 30* with inlet debris guard fitted. The variable stator swivel pins can be identified. *Fig 31* shows a halfcasing and mounting of the swivelling stators and actuating rings.

Rolls-Royce was first in applying this principle in the design of axial compressors and, although the idea was taken up by others around the world, Rolls-Royce did not use it themselves in a production engine until the much higher pressure ratio (16:1) of the V2500 was called for in the 1980s.

The Avon RA3 (100 series) was developed to give a dry thrust of around 8000 lbf. Beyond that, a new design of engine was required. At least 10000 lbf was going to be required for the Hunter Mk 6, Canberra PR9, Sud-Aviation Caravelle, English Electric Lightning, Saab Draken and Comet 3. There was now the requirement of designing a new axial jet engine of higher pressure ratio (7.5:1) and airflow (140 lb/sec).

This was an ideal opportunity to design a compressor with multiple stage variable stator blading, however, this would undoubtedly have been a major mechanical development task. At the level of compression ratio being considered, it had already been shown that the simpler solution of variable inlet guide vanes alone combined with progressively controlled interstage bleed was successful; this was adopted again for this new compressor.

Fig 30 General view of the variable stator rig compressor

Fig 31 View of half of the stator casing of the variable stator rig compressor

In addition to these two essential stall control features, the aerodynamic design of the first four stages of blading followed the design of the competing Armstrong Siddeley Sapphire turbojet which featured a graded increase in temperature rise per stage over the first four stages. The first stage temperature rise was reduced to 12°C as compared to 20°C of the Avon RA3. This reduction in first stage temperature rise lessened the severity of the stalling of the blading and resulted in lower sensitivity to surge of the compressor in the engine. The temperature rise per stage of the compressor was increased over these first four stages to 20°C at Stage 5 and for the remaining stages. These first four stages were matched to a new Rolls-Royce design for the last 11 stages making a total of 15 stages.

The new engine (*Fig 32*) was designated Avon RA14 (200 series) and it included an improved two-stage turbine and a tubo-annular combustion chamber with improved ducting from the compressor outlet to the flame tubes. From the beginning, the engine performed well attaining the design dry thrust of 9500 lbf in a matter of days and within a few weeks, by exploiting the advantages of the variable inlet guide vanes, was uprated to 10500 lbf. The Avon RA14 went on to be developed to a dry thrust of over 12000 lbf and 16000 lbf with reheat, becoming the most successful axial jet engine in the world until overtaken some seven years later by the General Electric J79 with multiple stage variable stator blading, allowing them to increase the cycle pressure ratio of the engine.

At the end of 1952, having considered both the two-shaft and single-shaft with variable stator concepts, General Electric chose the latter for their new advanced turbojet, J79 (*Fig 33*). The compressor had 17 stages, a variable inlet guide vane and six stages of variable stator and generated a pressure ratio of 12.9:1 and delivered 169 lb/sec of air (77 kg/sec). The engine flew for the first time in February 1956 in the Lockheed F104 Starfighter, entering squadron service in January 1958. The engine became a standard in the United States Air Force and will perhaps be best remembered as the powerplant in the McDonnell F4 Phantom. *Fig 34* shows the variable stator mechanism on the outside of the J79 compressor casing.

Having shown the powerful effect of inter-stage bleed flow on axial compressor off-design performance, the next Rolls-Royce axial jet engine, the RB80 Conway (*Fig 35*), exploited this to the full by designing the compressors to have permanent inter-stage bleed combined with the idea of two compressors in series driven by independent turbines on separate shafts. In the search for improved fuel economy, it was necessary to raise the pressure ratio and the turbine inlet temperature of the engine without increasing the velocity of the propulsive jet. This was necessary so that the propulsive efficiency of the engine in flight would be maintained or bettered. The permanent inter-stage bleed was the solution allowing the use of a four-

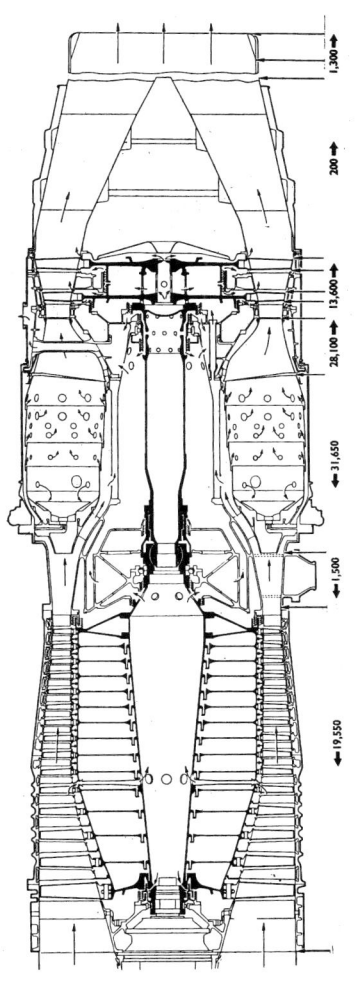

Fig 32 Cross-section of Rolls-Royce Avon RA14 axial turbojet

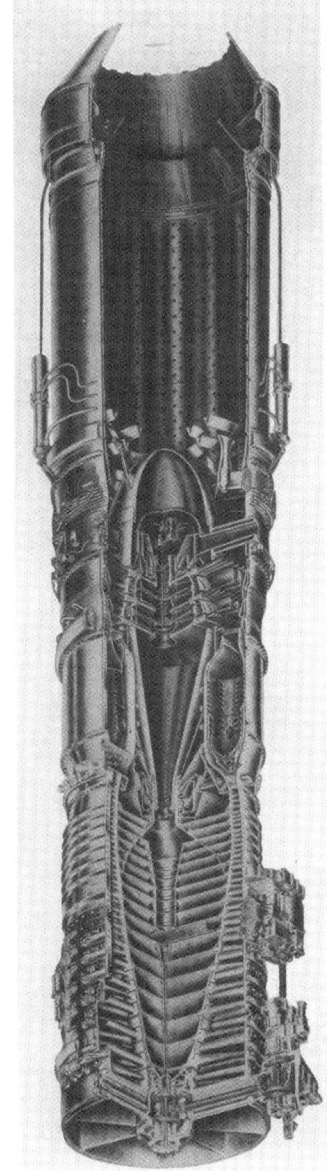

Fig 33 Cross-section of General Electric J79 axial turbojet engine

stage LP compressor giving a pressure ratio of 2.7:1 and an eight-stage HP compressor of 4.4:1 enabling the overall engine pressure ratio to be increased to 12:1 with substantially the same number of stages as the Avon turbojet. A further helpful feature is that with each compressor being driven by its own independent turbine on a separate shaft, the LP compressor slows down relative to the HP compressor at the lower rotational speeds. This has the effect of reducing the air incidence angles on to the blades in the LP compressor. This type of engine was initially known as a bypass turbojet and later, as the bypass ratios were increased to further improve propulsive efficiencies, this type of engine became known as a turbofan.

Later, in 1962, this same design concept was applied to the small RB172 bypass engine which became the Rolls-Royce Turbomeca Adour engine *(Fig 36)*. A two-stage LP compressor giving 2.0:1 and a five-stage HP compressor of 4.5:1 resulted in an overall compression ratio of 9.0:1. Half the air delivered by the LP compressor is directed around the engine to mix with the engine exhaust ahead of the propulsion nozzle, thus providing a permanent bleed to prevent the stalling of the LP compressor blading at low engine speeds.

The HP compressor and turbine operate very much as they would in a simple turbojet engine such as that illustrated in *Fig 3*. The pressure ratios

Fig 34 View of the variable stator vane mechanism on the compressor casing of the General Electric J79 axial turbojet.

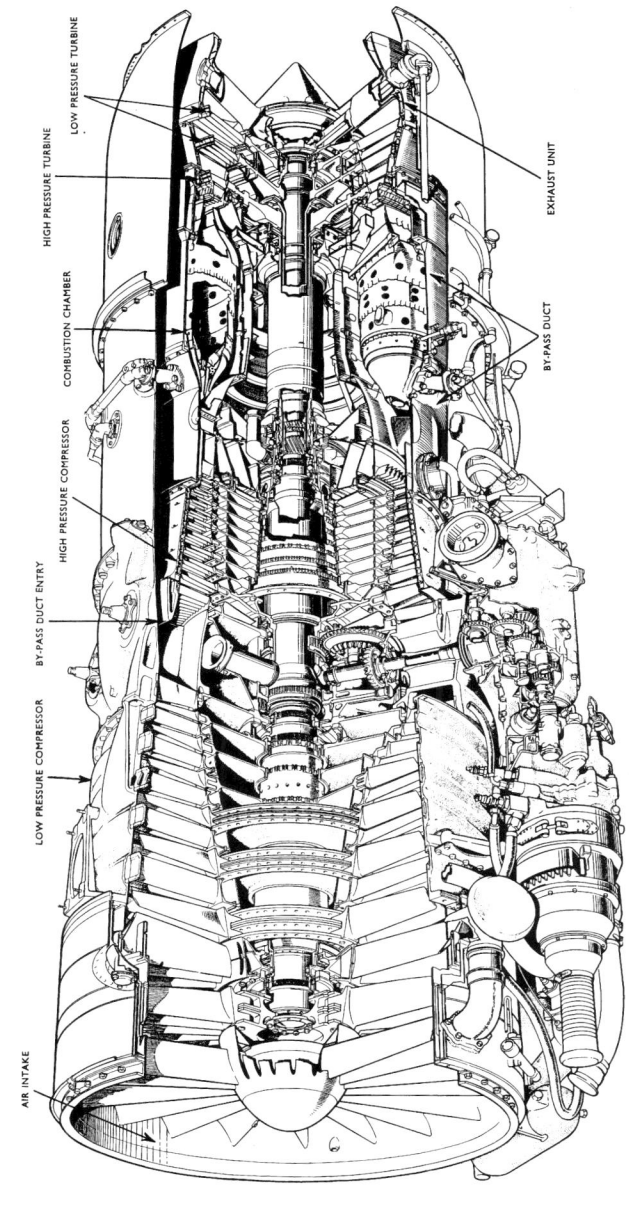

Fig 35 Sectioned drawing of Rolls-Royce Conway RCo12 bypass axial turbojet

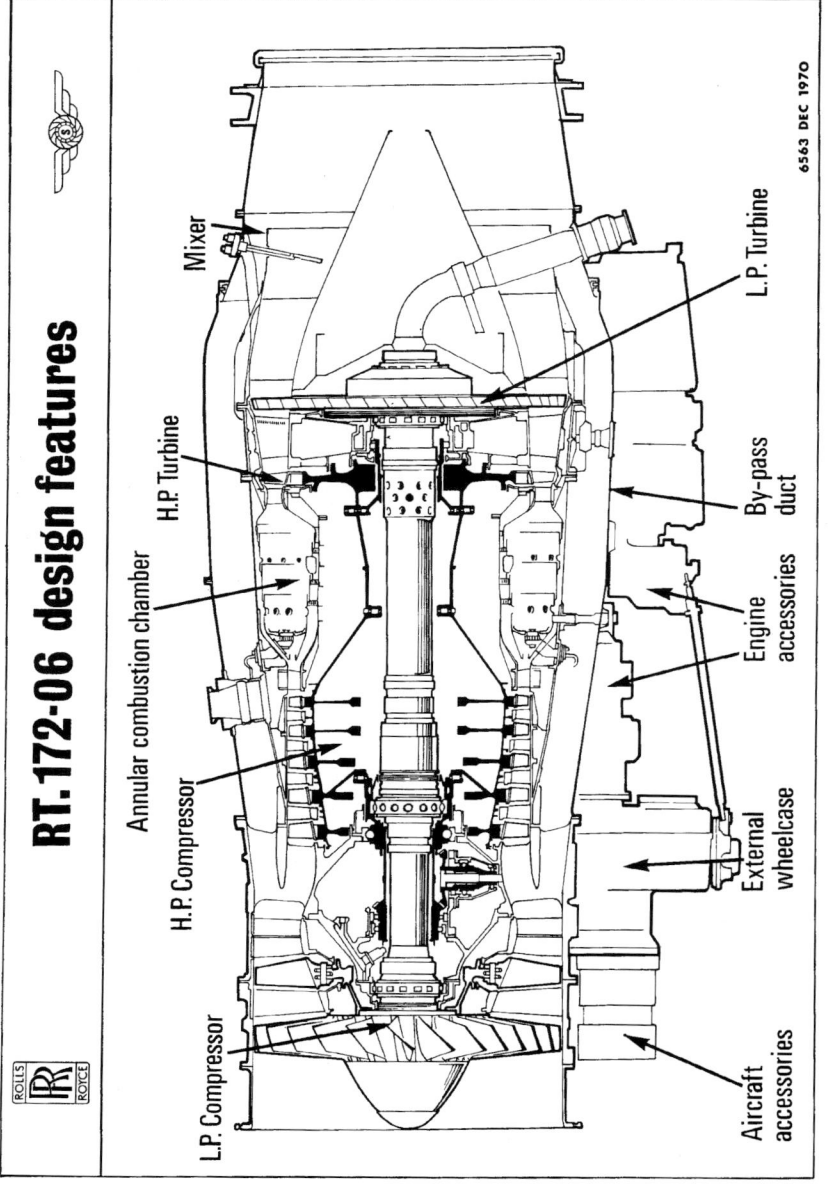

Fig 36 Cross-section of Rolls-Royce Turbomeca Adour turbofan

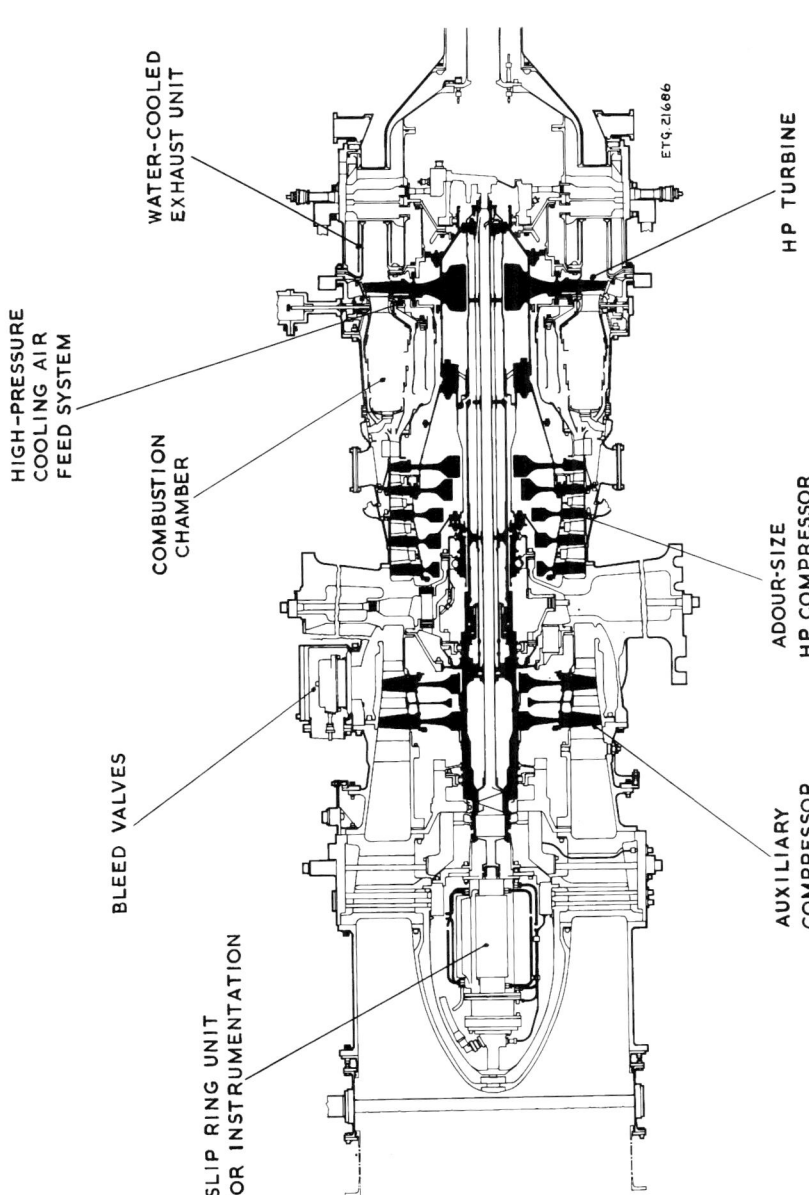

Fig 37 Cross-section of Rolls-Royce High Temperature Demonstrator Unit

are about the same and, as indicated in Chapter 3, the first stage would only be running in a stalled state at very low engine speeds where the adverse effects on performance, surge, vibration and blade stresses are much reduced. This Adour engine has been very successful in practice powering the Anglo-French Jaguar strike aircraft and the Hawk trainer/light fighter.

Another interesting engine project that required the compressor never to stall or surge was the High Temperature Demonstrator Unit (HTDU) specially designed and built (1968/70) to advance the design of high pressure, high temperature turbines. It could be called a turbine research engine. *Fig 37* is an illustration of the longitudinal section. The aim was to run new designs of air-cooled turbines at the highest possible temperatures (to 1800°K) at high pressure and high expansion ratios. For this work it was essential to have the most dependable compressors that would never surge or fail mechanically so that the vital turbine cooling development work would not be interrupted.

The design of the HTDU was based upon the high pressure (HP) system of the Adour engine because this had been well tested and was of proved reliability in aircraft service. To this were coupled two booster axial stages to raise the pressure ratio to 8.4:1. These were driven from the front of the Adour compressor and all driven by the high temperature air-cooled 'research' turbine. The two additional compressor stages were special to the HTDU designed to conservative stress levels for reliability and mounted on their own separate shaft and bearings. Thus, a long annular duct could be incorporated in the design between the two booster stages and the HP compressor. This duct was designed to include a generous size annular passage in the outer wall from which bleed air could be taken to a series of bleed valves. These were regulated according to a function of engine speed during the turbine tests to ensure that the two booster stages were never run with the blading stalled. The HTDU was the most highly instrumented engine in Rolls-Royce at that time. Turbine gas and blade metal temperatures were measured including blade vibration by strain gauges, all at very high temperatures. *Fig 38* is a photograph of the HTDU fully instrumented for test.

The tests were a mixture of endurance running at the design speed and temperature, and cyclic tests which required the engine to be accelerated from idling to full speed and back according to a time/speed schedule. It was to ensure surge free running of the compressors during the cyclic tests in particular that determined the design layout described. The whole system worked remarkably well.

With the compressor configuration shown it is possible to plot the characteristics of the two booster stages and the five HP stages separately even though they are on the same shaft and rotate at the same speed. *Fig 39* shows the HP compressor characteristics, the design point pressure ratio being 4.25:1. The different compressor operating lines shown are for four values of intake ram ratio (R_{01} =1.0; 1.2; 1.5; 2.0 respectively). The intake ram ratio is the ratio of the engine inlet total pressure over the ambient static pressure to which the engine nozzle exhausts. It increases with increasing aircraft speed and is the pressure ratio by which the intake to the HTDU is raised by the test plant to increase the 'research' turbine inlet pressure to a level of 16 atmospheres. At a ram ratio of 1.0 the compressor working line is satisfactorily well below the compressor surge line, and at higher ram ratios at the low engine speeds it is much further away. This movement occurs at the low speeds because the turbine is not choked at these conditions (throat Mach No < 1.0). At the higher speeds where the turbine is choked (throat Mach No = 1.0) the compressor working line position is little affected.

Fig 38 Fully instrumented HTDU ready for test

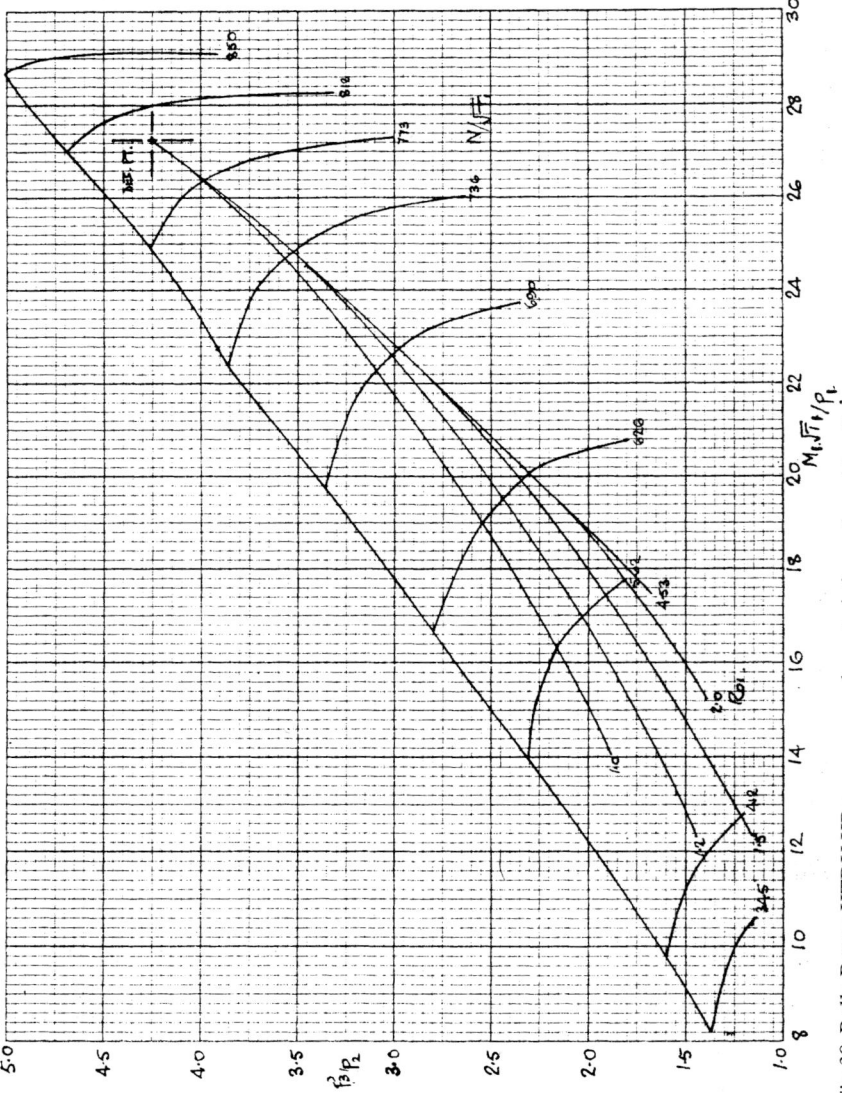

Fig 39 Rolls-Royce HTDU HP compressor characteristics and working lines

Fig 40 shows the low pressure (LP) or 2 Stage booster characteristics, the design point pressure ratio being 1.97:1 giving the product 8.4:1 for the combined 7 stages. From Chapter 3 of this paper it would be expected that a compressor with a design pressure ratio of 8.4:1 would cause the front stages of the compressor to be stalled at the lower speeds initiating surge. Reference to *Fig 40* shows this to be the case. The compressor working line intersects the surge line at a pressure ratio of 1.65:1, or at 83% of the booster compressor design speed. By allowing bleed air to escape between the two compressors as described (*Fig 37*) the compressor operating line on the booster compressor characteristics is moved to higher flows on to the stable region of the flow-pressure ratio curves, 10% bleed being shown in *Fig 41* and 15% bleed in *Fig 42*. It will be understood that for the cyclic engine tests the bleed flow has to be variably controlled so that at the higher speeds the bleed is closed off to enable the booster stages to deliver the full design pressure ratios to the HP compressor at the high speeds.

An especially interesting example of a high pressure ratio turbofan engine, the V2500, the design of which had to be changed to overcome compressor stall, is shown in *Fig 43*. It was an ambitious design in which the high pressure compressor had a pressure ratio of 20:1 with four variable angle stators to overcome the stalling of the first four stages at the lower engine speeds. Unfortunately the angles through which the stator blades had to be rotated to reduce the incidence angles were so large that the diffusing passages between the blades tried to impose such a high rate of diffusion in the flow that flow separation and turbulence developed. Excessive rotation of the blades to reduce the incident air angle had eventually initiated a new stall at the much larger angles due to the excessive blade passage area diffusion rate thus introduced! This demonstrated that there are angles to which movable stator blades of a particular design must not be allowed to exceed (cf *Fig 16*). The situation was very serious because after starting the engine it would not accelerate due to surge.

The only solution was to reduce the pressure ratio of the compressor to 16:1 in the same 10 stages and to restore the overall pressure ratio of the engine by increasing the number of booster stages behind the fan from one in *Fig 43* to three in *Fig 44* and provide an extra 5th stage of variable stator. These design changes were fundamental to the successful functioning of this engine and represented a very considerable effort both in design and development to meet the engine performance commitment at the planned date. Thus, compressor stall and surge had again become a major stumbling block to progress as it had been 40 years earlier in the case of the Avon turbojet.

Over this period, there have been advances in theory and methods of calculating the three-dimensional flow in axial compressors together with the

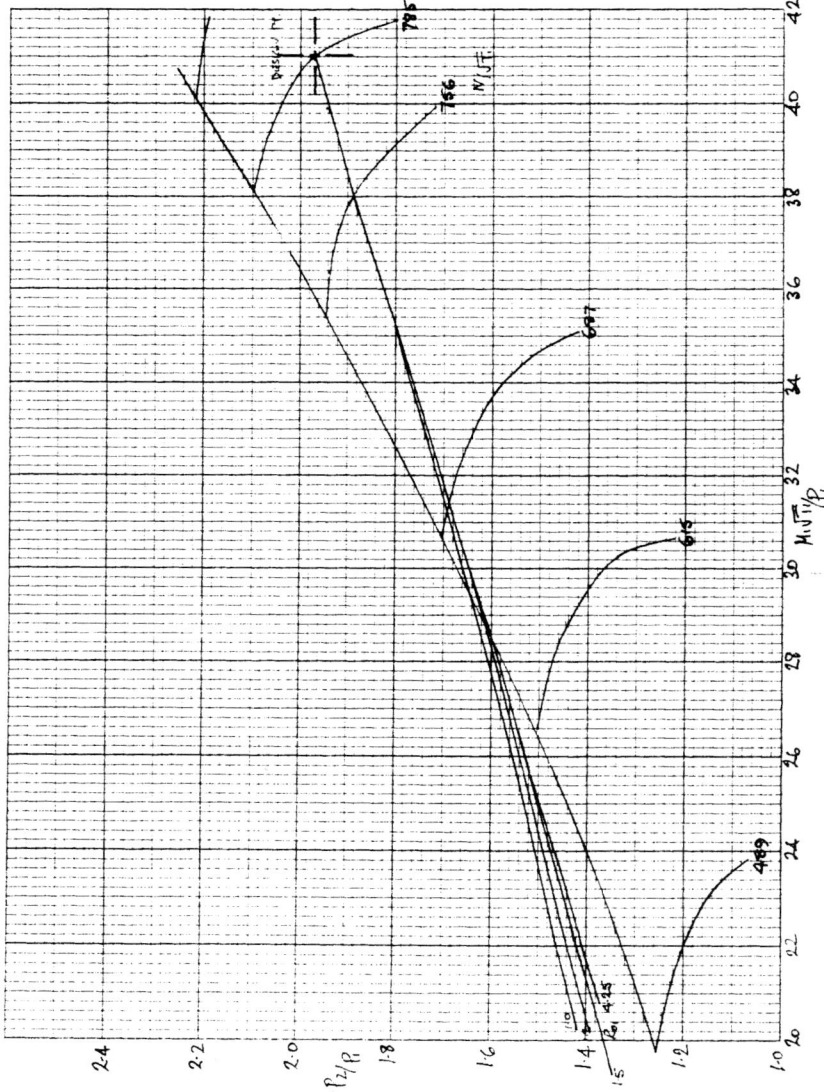

Fig 40 Rolls-Royce HTDU auxiliary compressor (booster - LP) characteristics and working lines

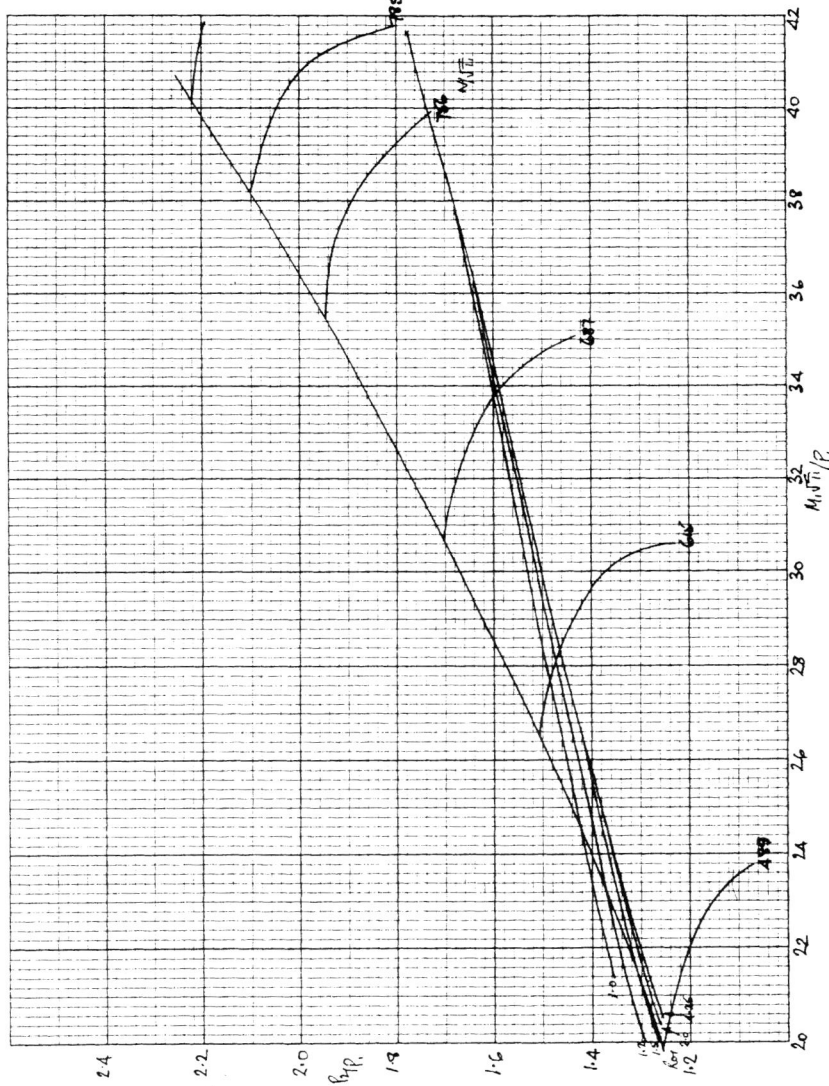

Fig 41 Rolls-Royce HTDU auxiliary compressor (booster - LP) characteristics and working lines with 10% bleed

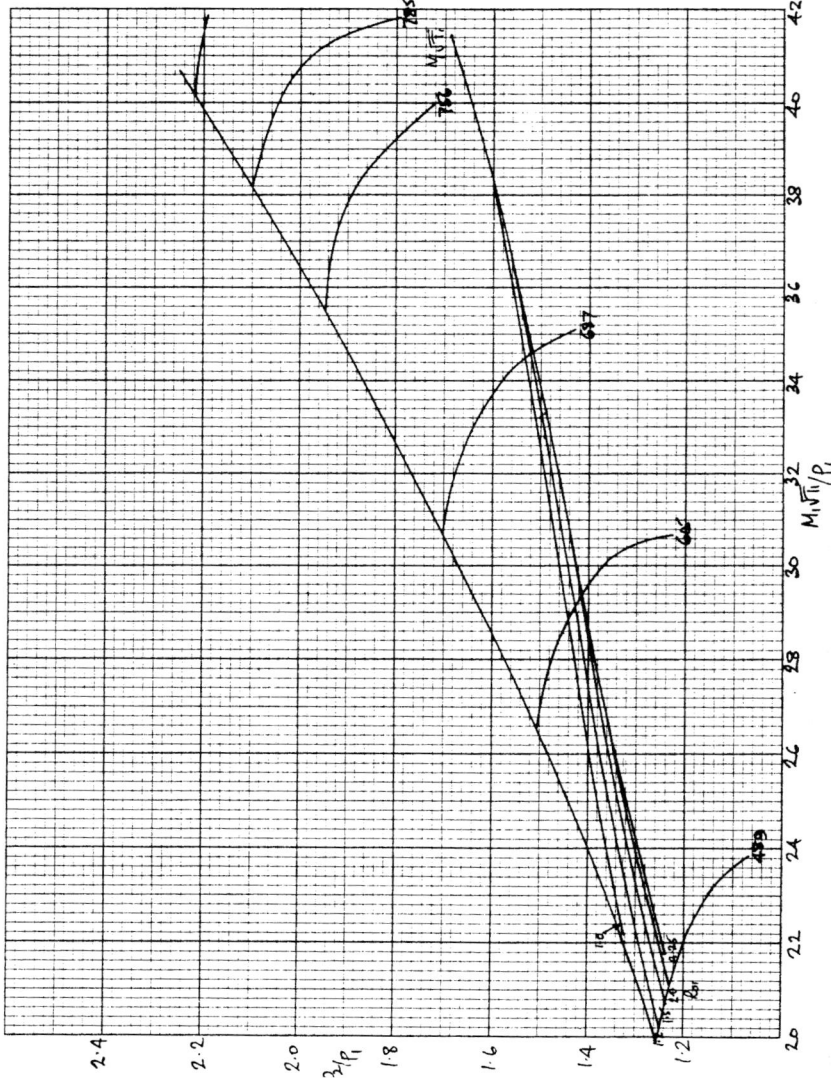

Fig 42 Rolls-Royce HTDU auxiliary compressor (booster - LP) characteristics and working lines with 15% bleed

development of blading capable of sustaining higher diffusion coefficients at higher Mach No.

The consequence of this has been that whereas the Avon axial compressor needed 12 stages to produce a pressure ratio of 6.3:1 at the design condition, the V2500 axial compressor 40 years later required only 10 stages to produce 16:1 at the design condition.

By the 'design condition' here is meant the requirement of a flow rate, a pressure ratio and a declared minimum efficiency. Although demanding, this is a small part of the total requirement which must include in addition pressure ratio and flow margins to ensure smooth compressor operation in engine acceleration and deceleration running modes. Unfortunately, so far, there are not any reliable theories to prescribe these margins which require the prediction of the compressor surge pressure ratio-flow rate relationship with variation of rotational speed. The flow fields are too complex to be prescribed in these regions. The prediction of the 'surge line' is largely empirical based upon the correlation of a large number of experiments and compressor tests.

It can be expected that developments of aerodynamic theory and computational fluid dynamics will contribute to small increases (1 to 2%) in compressor efficiency at the design condition. However, it is unlikely that the lower speed off-design performance will benefit from more sophisticated aerodynamic theories when the effect of varying the stagger angle of stator blading to postpone stall creates unpredictable flow distribution across the annulus between the variable stagger angle stator blading.

The advances in axial compressor performance have been the major contributing factor in the design of efficient turbofan engines. In this, imagination and innovative mechanical design have been at least as important as theory, aerodynamic design and development. It is important for students in engineering to understand this. With the application of CFD in all fluid flow problems at great expense, it is as important to know what it is likely to achieve as to know what it cannot achieve. Often an experiment is more productive and less expensive.

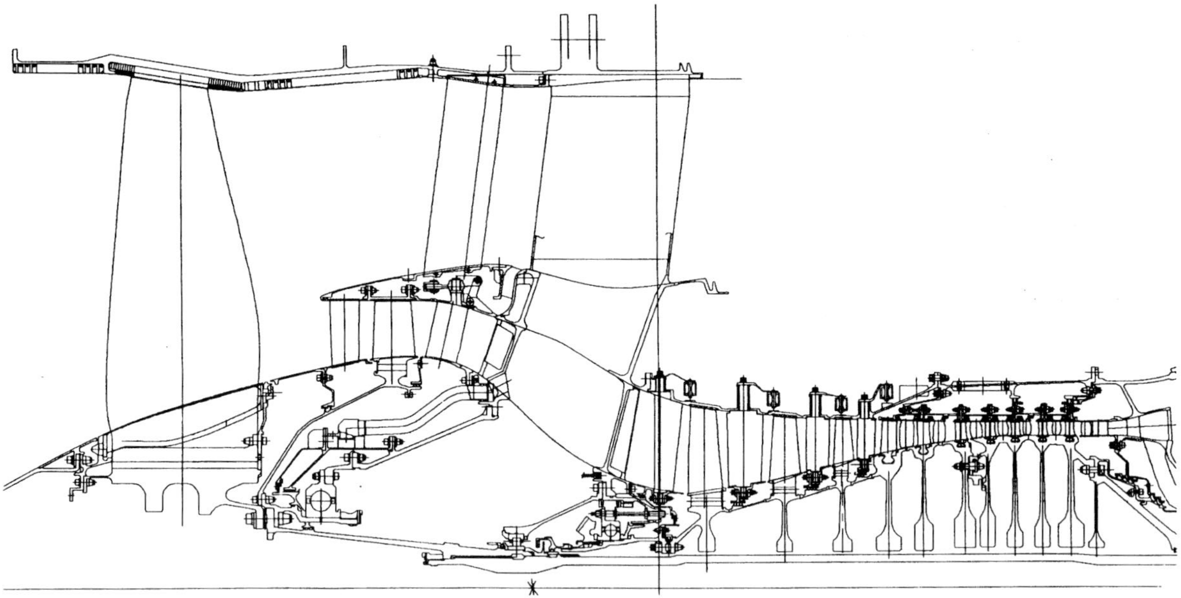

Fig 43 IAE V2500 original design standard with one booster stage and four variables on the HP compressor

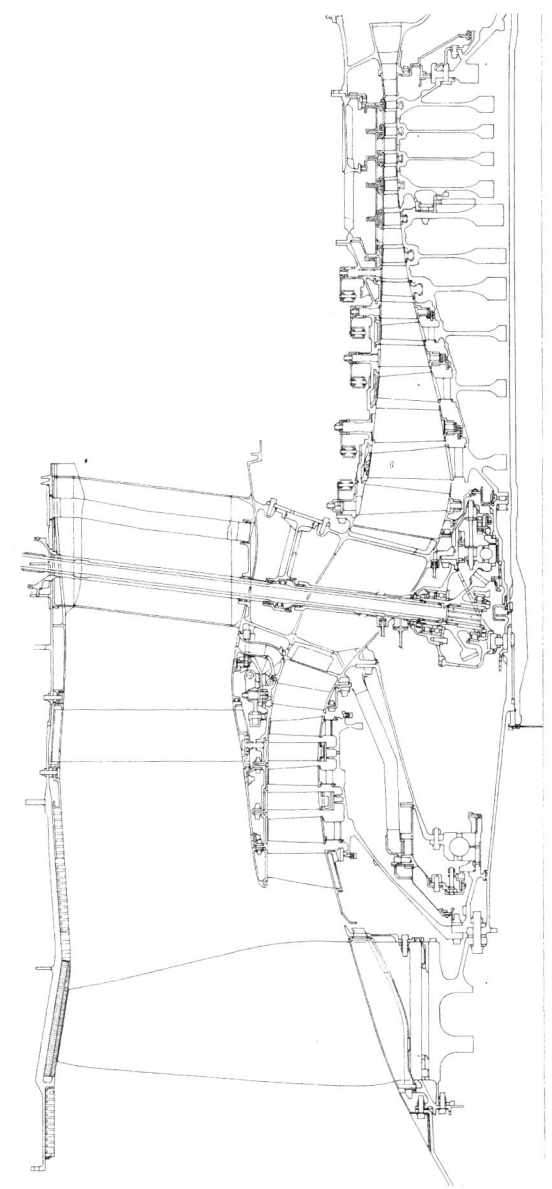

Fig 44 IAE V2500 certification standard with three booster stages and five stanges of variables on the HP compressor

CHAPTER EIGHT

Summary

The fundamental cause of the stalling of blading in the axial compressors at off-design conditions has been explained, albeit by drastically simplified data and procedures. Anyone concerned with the design or operation of engines or industrial power plant systems in which axial compressors are used cannot afford to be totally ignorant of the off-design performance characteristics of axial compressors.

It has been stated that compressor flow and delivery pressure become dangerously unstable after blading has been caused to operate in the stalled state. This is broadly true, but the degree to which the blading has to be stalled for the phenomenon of surge to occur varies with the aerodynamic design of the blading. Blading which produces a high pressure rise per stage is more critical to the onset of surge than blading that produces a low pressure rise per stage. The sensitivity to surge relates to the deflection of the air that the blading is designed to produce, and the incident Mach No of the flow on to the aerofoils. This report states that surge takes place but does not explain how it develops. That is a subject not fully understood even today and theoretical treatments have not so far yielded any very useful practical guides.

The only really useful design approach to avoiding surge is to design compressors in which the blading does not have to run in stalled conditions. This can be done and several practical examples of machines are included in the report.

It is thought that even with the simplified approach to the problem adopted in this paper, the engineering student should be able to grasp the fundamentals and become involved in the interesting aerodynamically related mechanical design solutions. There is still scope for aero/mechanical design ingenuity in the application of axial compressors to all kinds of engines and industrial plant generally.

The Historical Series is published as a joint initiative by the Rolls-Royce Heritage Trust and The Sir Henry Royce Memorial Foundation.

Also published in the series:

No.1	Rolls-Royce – the formative years 1906-1939 Alec Harvey-Bailey RRHT 2nd edition 1983 (out of print)	
No.2	The Merlin in perspective – the combat years Alec Harvey-Bailey, RRHT 4th edition 1995	
No.3	Rolls-Royce – the pursuit of excellence Alec Harvey-Bailey and Mike Evans, HRMF 1984	
No.4	In the beginning – the Manchester origins of Rolls-Royce Mike Evans, RRHT 1984	
No.5	Rolls-Royce – the Derby Bentleys Alec Harvey-Bailey, HRMF 1985	
No.6	The early days of Rolls-Royce – and the Montagu family Lord Montagu of Beaulieu, RRHT 1986	
No.7	Rolls-Royce – Hives, the quiet tiger Alec Harvey-Bailey, HRMF 1985	
No.8	Rolls-Royce – Twenty to Wraith Alec Harvey-Bailey, HRMF 1986	
No.9	Rolls-Royce and the Mustang David Birch, RRHT 1987	
No.10	From Gipsy to Gem with diversions, 1926-1986 Peter Stokes, RRHT 1987	
No.11	Armstrong Siddeley – the Parkside story, 1896-1939 Ray Cook, RRHT 1989	
No.12	Henry Royce – mechanic Donald Bastow, RRHT 1989	
No.14	Rolls-Royce – the sons of Martha Alec Harvey-Bailey, HRMF 1989	
No.15	Olympus – the first forty years Alan Baxter, RRHT 1990	
No.16	Rolls-Royce piston aero engines – a designer remembers A A Rubbra, RRHT 1990	
No.17	Charlie Rolls – pioneer aviator Gordon Bruce, RRHT 1990	

No.18	The Rolls-Royce Dart – pioneering turboprop Roy Heathcote, RRHT 1992
No.19	The Merlin 100 series – the ultimate military development Alec Harvey-Bailey and Dave Piggott, RRHT 1993
No.20	Rolls-Royce – Hives' turbulent barons Alec Harvey-Bailey, HRMF 1992
No.21	The Rolls-Royce Crecy Nahum, Foster-Pegg, Birch, RRHT 1994
No.22	Vikings at Waterloo – the wartime work on the Whittle jet engine by the Rover Company David S Brooks, RRHT 1997
No.23	Rolls-Royce – the first cars from Crewe Ken Lea, RRHT 1997
No.24	The Rolls-Royce Tyne L Haworth, RRHT 1998
No.25	A View of Ansty David E Williams, RRHT 1998
No.26	Fedden – the life of Sir Roy Fedden Bill Gunston OBE FRAeS, RRHT 1998
No.27	Lord Northcliffe – and the early years of Rolls-Royce Hugh Driver, RREC 1998
Special	Sectioned drawings of piston aero engines L Jones, 1995

Technical Series:

No.1	Rolls-Royce and the Rateau Patents H Pearson, RRHT 1989
No.2	The vital spark! The development of aero engine sparking plugs K Gough, RRHT 1991
No.3	The performance of a supercharged aero engine S Hooker, H Reed and A Yarker, RRHT 1997

Books are available from:
Rolls-Royce Heritage Trust, Rolls-Royce plc, Moor Lane, PO Box 31, Derby DE24 8BJ